safetyFUNdamentals™

77 Games and Activities to Make Training Great!

Linda M. Tapp, ALCM, CSP

Ennismore Publishing

SafetyFUNdamentals: 77 Games and Activities to Make Training Great!

By Linda M. Tapp, ALCM, CSP

Published by:

SAN 850–6701
Ennismore Publishing
1999 Route 70 East, Suite 13
Cherry Hill, NJ 0003 U.S.A.
info@ennismorepublishing.com
www.ennismorepublishing.com

Printed in the United States of America.

Copyright © 2006 by SafetyFUNdamentals™ and Linda Tapp

All rights reserved. No part of this book may be reproduced or transmitted in any form or by any means, electronic or mechanical without written permission from the author, except for the sample games included in the Appendix which may be reproduced without permission.

ISBN: 0-9779324-2-7

Limit of Liability/Disclaimer of Warranty:
While the Publisher and Author have used their best efforts in preparing this book, they make no representations or warranties with respect to accuracy and completeness of the content of this book and specifically disclaim any implied warranties of merchantability or fitness for a particular purpose. No warranty may be created or extended by sales representatives or written sales materials. The advice, strategies, games and activities contained herein may not be suitable for your situation or facility. While these games and activities are appropriate in most situations and for most audiences, and safe if followed as described, neither the author nor the Publisher shall be liable or responsible for any loss, injury, or damage allegedly arising from any of the information, instructions or suggestions in this book. The games and activities provided in this book are not meant to take the place of a comprehensive site specific safety and health program. You should consult with a professional where appropriate. Neither the publisher nor author shall be liable for any loss of profits or any other commercial damages, including but not limited to special, incidental, consequential, or other damages.

Library of Congress Control Number: 2006906682

Copyright © 2006 SafetyFUNdamentals™

SafetyFUNdamentals: 77 Games and Activities to Make Training Great!

is dedicated to all safety and health professionals who really want to make a difference in the lives of the workers they have the privilege to train. Your energy and commitment is more valuable than you will ever know.

Copyright © 2006 SafetyFUNdamentals™

Copyright © 2006 SafetyFUNdamentals™

Table of Contents

Copyright © 2006 SafetyFUNdamentals™

Copyright © 2006 SafetyFUNdamentals™

Copyright © 2006 SafetyFUNdamentals™

Copyright © 2006 SafetyFUNdamentals™

Chapter 1

Introduction

Congratulations! You have decided to use games and other interactive activities in your safety training classes. As you probably know from first hand experience, there are many, many boring safety training classes being given every day across the world. If you are reading this, you probably decided that you do not want to be leading that type of class. Safety training is just too important! Whether or not your trainees "get it" can mean the difference between someone having a serious accident (or worse) and someone knowing when and how to do the right thing and work safely.

Safety Training is just too important!

Even if the video you are showing or PowerPoint® slides you rely on are effective, the probability that your trainees will remember much of what you say in even a few weeks is low if they have not actively participated in the class. Games and interactive activities are one of the best ways to do just this.

Open Your Mind!

Before you proceed, promise yourself that you will read through the following 77 activities with an open mind. Many individuals asked to present safety training are not trained facilitators or trainers so using these types of activities may take them out of their comfort zone. Does this sound like you? Go for it anyway! You can have a huge impact on the safety of those working in your plant and at home if you just give it a try.

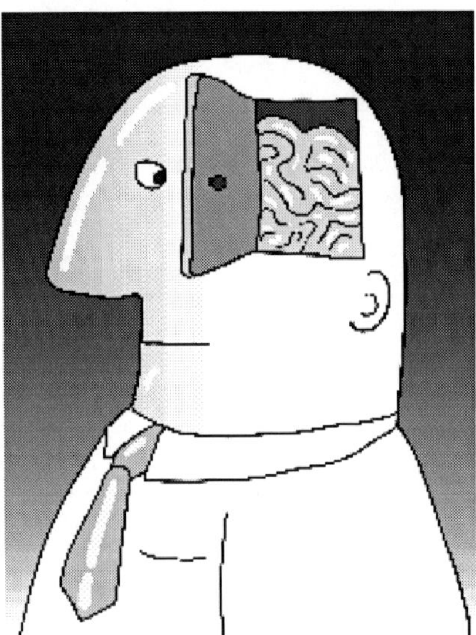

What about the trainees? Will they think you are crazy? You will be surprised. They are as tired of sitting through boring safety training classes as you are giving them. At first you may get a few complainers but that can and will happen with almost any class. Soon, they will be enjoying themselves too.

> **There exist limitless opportunities in every industry. Where there is an open mind, there will always be a frontier.**
>
> **- Charles F. Kettering**

Copyright © 2006 SafetyFUNdamentals™

A Worthwhile Risk

Safety and health professionals are generally a risk-adverse bunch but when using interactive training activities, risk is good! These are "Right Risks" as defined by Bill Treasurer in his book titled "Right Risk: 10 Powerful Principles for Taking Giant Leaps with Your Life." Mr. Treasurer says that right risks are often those that offer the potential to fill our gaps. Mr. Treasurer goes onto say, "Right Risks are deliberate, focused, and rich with meaning." Being the first in your organization to try using games and other interactive activities in your training classes may seem like a risk but these activities can help to fill the gaps that have existed in safety training for so long.

Many people in all industries are uncomfortable when having to stand up in front of an audience although it is something they probably need to do rather often. If they are also being asked to try something out of the ordinary like an interactive safety game, this may take a bit more convincing.

Many a man is praised for his reserve and so-called shyness when he is simply too proud to risk making a fool of himself.

- Joseph Priestley

Why do safety professionals do anything they feel uncomfortable with? Because their actions could make the difference between someone living and someone dying. Although we may not think about safety training as having the same power - it does. Whether or not someone pays attention and actually learns and remembers what you are telling them could have a profound impact on their personal safety. Trying to do the best we possibly can is a risk we have to take.

On a lighter note, once the safety trainer gets used to facilitating games and activities, he or she will see that these aren't so risky at all. With practice comes confidence and by doing these games and activities, the risk will seem to disappear. Trainees will not only enjoy class more, they will remember more and you will see safety training become what it was meant to be.

Copyright © 2006 SafetyFUNdamentals™

A Word about the Terms Used in this Book

Trainees vs. Students

As Dr. Scott Geller said in his excellent article titled "Are You Training or Educating ?" (Industrial Hygiene and Safety News, May 2000), we often use the tems safety education and safety training interchangeably although we know that providing education and training are not the same things. In some instances, we need to provide education and at other times, safety training is more important. Sometimes, a combination of both is necessary. The training activities in this book are most effective after the trainees have had the related education. Along these lines, you will see that the class attendees are referred to as "trainees" and not "students," although in some of the classes you offer they may be both.

For example, in a typical forklift training class, the student/trainee first sits through a classroom presentation on safe forklift operation and later, he or she must demonstrate their skill in operating a forklift safely (and if they have never driven a forklift in their life, they will be shown where the controls are and the basic operation). Without the education portion of this class, could the operator demonstrate safe use? Alternatively, without having to demonstrate safe use, would you be comfortable letting that operator drive a forklift? In most cases, both education and training are necessary. Make sure you are providing both.

Copyright © 2006 SafetyFUNdamentals™

Accelerated Learning

Accelerated learning principles, applied to safety training, can be very powerful tools - especially when you consider the topics that safety trainers often try to teach. These are often life and death issues and the absorption and retention of this information is extremely important. Traditional safety training usually involves showing a video or PowerPoint slides and then discussing it as it relates to the company's program and policies. Accelerated learning is much different. Accelerated learning is focused on the results – not the methods. Whatever learning tools work to increase and enhance learning can be called accelerated learning methods. Many safety professionals spend a good portion of their work week training others. Accelerated learning principles can help you spend less time creating training programs and can help trainees learn faster and remember more. Accelerated learning principles applied to safety training can be a win-win situation for everyone.

What is accelerated learning? Let's first discuss what it is not. Accelerated learning, although it often involves games, imagery and sometimes even music, is not just a bunch of activities to do to pass the time. Everything in an accelerated learning class is focused on the results and not the materials or activities themselves. For example, a safety trainer announces that everyone is going to get up play Twister® with the goal of getting everyone relaxed and "ready to learn." While this might seem like a fun icebreaker, the Twister® game is fun only for the sake of being fun. None of the information to be covered in this training is being reinforced by this activity. In an accelerated learning class, we might also play games but the games have a different focus – on the results, instead of the activity. For example, a BINGO game could be played where instead of a number and letter being called out, a clue is called. Instead of numbers being on the BINGO card, answers directly related to the training content are listed. This game is still fun, and gets everyone involved, but includes accelerated learning principles, especially when you have small teams work on each BINGO card instead of individually. A sample BINGO game used in an office ergonomics class as well as the clue sheet is provided in Appendix A.

Copyright © 2006 SafetyFUNdamentals™

The major principles of accelerated learning, as explained in Dave Meier's *Accelerated Learning Handbook* ©2000, can be simply stated as:

- Complete trainee involvement
- Dynamic creation of knowledge and applicable job skills
- A group learning effort whereby trainees work together
- Interactive learning instead of or in addition to lecture format

To review how these principles apply to safety and health training, consider the following. Learning should involve the whole mind and body. Learning does not just take place in the head. In Mr. Meier's book, he states that training should be "SAVI" which basically means that training should apply to all of the senses (somatic, auditory, visual and intellectual).

Training needs to be created by the trainee and not necessarily given to them. When learners learn, they are creating new meanings and associations for themselves. One thing that good safety training always tries to accomplish is to get the trainee to understand how the training applies to them and their job in a realistic way. They should know how this training affects their efforts to maintain a safe workplace.

Good learning is social and we can learn much more by learning with our peers than we can by ourselves. Most of us probably learned in an environment that involved a great amount of individual competition. When trainees work together in teams, learning can be greatly enhanced.

Learning takes place by absorbing many things at once and not one fact at a time. The brain works best when it is challenged and asked to do many things at the same time. If you ask someone to sit still and just look at a bunch of slides or listen to a speaker, learning will not be as great as if the same material was presented along with an activity that would relate to the material at hand. Activity, or doing the work itself, enhances learning.

Copyright © 2006 SafetyFUNdamentals™

We have all heard the famous Confucius quote "I hear and I forget. I see and I remember. I do and I understand." In almost every area of life, we learn by doing. For example, you can have someone read endless safety policies and manuals and watch videos about safely driving a forklift but until the person actually gets on and drives it, real learning will not take place.

Positive emotions are also very important in enhancing learning. If someone is sitting endlessly in a lecture, they probably won't have positive emotions for very long. Activity can help with these positive emotions. If someone is stressed or bored or angry, their learning will be inhibited. If learning is positive, relaxed and engaging, learning will be increased. Activities can help to keep the trainee engaged.

Also related to accelerated learning activity is the use of the "image brain". Our nervous system is more of an image processor than a word processor so any time we can take safety concepts and make them visual, the trainee is more likely to learn. Activities are a great way to turn the safety concepts we must teach into actual images that will help the trainee learn more and faster.

Finally, safety training classes that incorporate accelerated learning principles can be designed in a fraction of the time it takes to design typical classes. When we put together a training class, we often spend a great deal of time on the PowerPoint presentation, the hand-outs and making copies of related materials like sections of OSHA standards. When you realize that people learn much more, and much faster from experiences combined with feedback, you will begin to see how accelerated learning principles applied to safety can save you a lot of time.

You may not be able to eliminate the PowerPoint format entirely but the written materials will not be the focal point and they will not be the most important part of your training. Instead, the slides can help to initiate, guide and support the experiences that are used in the class.

Copyright © 2006 SafetyFUNdamentals™

Involving All Senses

It is important to point out that people will not automatically learn more because they are standing up and moving, but if you combine physical movement with intellectual activity and use all of the senses, this can have a profound effect on learning. The trainee does not need to be active 100% of the time but when interactive class activities are mixed with lecture or other typical presentation methods, retention will increase. If possible, all of the senses should be used since people learn in different ways. Some ideas for using some or all of the senses, along with intellectual activity, are described below. In this book, use of the senses is referred to as "body work," "sound work," "sight work," and "brain work."

"Body Work" in training means adding activities that include our abilities to touch and feel. "Body work" is often known as hands-on training. In safety, there are many opportunities for trainees to actually try out what you are telling them. Some ways that you can use "body work" in a safety training class include:

- Asking them to act out how something works (such as having them perform a LOTO procedure)
- Asking them to simulate how a structure or function in the human body works (In a respiratory protection class, you can give teams some basic tools and gadgets and ask them to show why particles of different sizes deposit in different parts of the respiratory system)
- Having them act out a communication process (ask them to act out giving feedback to a co-worker when an unsafe act was committed)
- Asking them to create large pictograms (this can easily be used with any safety and health topic)
- Having trainees talk about an experience and reflect on it (ask the trainees to try a new procedure and then report back to the class at a future training class)
- Completing a project that requires physical activity (have the trainees go out into the workplace and perform an activity related to the class and then return)
- Doing an active learning exercise such as a learning game (games such as crossword puzzles and BINGO can be great

Copyright © 2006 SafetyFUNdamentals™

learning tools for any safety and health topic as long as they are intellectually challenging enough)
- Taking a field trip out into the plant and then when trainees come back, have them write, draw or talk about what they learned (ask trainees to go out and inspect the workplace for a particular type of hazard)
- Interviewing people outside the class (have trainees complete interviews of coworkers on a particular safety topic and report back at the next class)

"Sound Work" in training means adding activities that include listening. Our ears are continually capturing all types of information and processing and storing it without us even realizing it. When we make our own sounds by talking about this information, it has even more staying power. All learners, especially auditory ones, learn by sounds. These sounds could be someone talking to them as in a lecture, reading out loud, talking with co-workers, or remembering jingles or rhymes.

Some ways to incorporate more "sound work" into your safety training classes include:
- Having trainees talk about what they are learning (this will work for any safety training class)
- Asking them to talk about an experience (this is especially effective if someone had a close call or actual incident they can share with the class)
- Asking them to read out loud or even act something out while reading it (you can ask trainees to take turns reading slides or hand-outs)
- Instructing trainees to talk through a problem and think out loud while they are doing some of the physical activities described. If they are sent out into the plant on an inspection, instruct them to "think out loud" about what they see good and what they see wrong during their tour
- Asking them to read something and then paraphrase it for the class (this can be very effective when it is important to go over the actual wording of an OSHA standard)
- Telling the class stories that have the learning concepts imbedded in the material (you can tell stories of your own or that you heard from others as long as they relate to the topic you are teaching)

Copyright © 2006 SafetyFUNdamentals™

- Having the trainees pair up and tell each other what they just learned and how they are going to apply it (this works well for any topic area)
- Asking the trainees to create a rap, rhyme or auditory mnemonic out of what they are learning (this also works well for any topic area but remember to have the class work in teams)

"Sight Work" in training means adding activities that involve seeing or visualizing. Visual acuity, or the amount of visual perception, is present in everyone. Like auditory sensitivity, visual acuity is more pronounced in some people than in others. Visual learners need to "see" what you are talking about. In this book, having trainees learn through their vision is referred to as "sight work." People who learn best through "sight work" learn best when they see real world examples, idea maps, pictures and images. Sometimes these people learn even better when they are asked to create their own images such as pictograms or mind maps. Some people also learn very well when they are told to observe a real world situation and then think and talk about it while drawing out the processes, principles or meanings.

To incorporate "sight work" into your safety training, you can include:

- Vivid presentation graphics (look for some real photos of your workplace – forget the boring clipart)
- Descriptive language such as metaphors and analogies (think of something that is directly related to your training that will help trainees remember)
- 3-D objects (bring actual objects into the class and if they are too big or heavy, have the class travel to the object)
- Vivid stories (as described above, stories can be your own or others but make sure you describe details and a very clear picture in order to capture the attention of the visual learner)
- Instructions to develop their own pictograms (this works well with any safety topic)
- Permission for the class to go out into the plant to do observations and then have them report back (you can provide a digital camera to the teams and have them take photos of what they see or you can ask them to sketch it)

Copyright © 2006 SafetyFUNdamentals™

- Colorful decorations and peripherals around the training room (boring training rooms often lead to boring classes – try to make the room fun)
- Mental exercises, such as visualization (ask the class to close their eyes and imagine themselves doing something they just learned about)

"Brain Work" is not academic work like experienced at school nor does it mean that someone must be smart. It refers to the process that is taking place in the trainee's head when he or she connects the material just learned with experiences that occurred in the past, or when they make plans for using the information just learned in the future. Basically, it is the process the learner uses to make sense of their new knowledge. Games and activities must always include this intellectual side of the trainee. It is extremely important not to leave this part of learning out when you are including "body work," "sound work," and "sight work." Without "brain work" training can seem frivolous and silly and even a waste of time. To include "brain work" in your safety training class, consider asking trainees to:

- Problem solve
- Analyze experiences
- Plan an activity
- Condense or interpret information
- Make up questions (perhaps even make up their own test questions)

Accelerated learning principles used in safety training classes can make training more effective. These principles can also decrease the amount of time necessary to create training materials as well as decrease the amount of times that attendees must be in class in order to learn and retain the information. Although accelerated learning principles may be very different from the way we usually learn as adults, these principles can greatly improve your safety training efforts.

Copyright © 2006 SafetyFUNdamentals™

Training is everything. The peach was once a bitter almond; cauliflower is nothing but cabbage with a college education.

- Mark Twain

Education is what survives when what has been learned has been forgotten.

- B. F. Skinner

If little else, the brain is an educational toy.

- Tom Robbins

Copyright © 2006 SafetyFUNdamentals™

What Makes a Game Effective?

What makes a game effective? Safety training games should incorporate a combination of chance and skill whenever possible. If too much chance is involved, the game becomes nothing more than a waste of time and a mindless activity. Imagine tossing dice for twenty minutes straight just for the purpose of trying to get the highest number. That would be boring. If a game involves too much skill, it quickly becomes a quiz and although quizzes have their place in safety training, you do not want a safety training game to turn into a test.

Effective safety training games should also be flexible and easily adaptable. If you find that the activity is too challenging for a particular class, you should be able to quickly modify it based on the abilities of the class you are currently training. Good safety training games should also be flexible enough so that you can do them in a variety of locations.

Is your game easily understood? If there is any confusion at all as to what the teams should be doing after you give your instructions, you will lose the team and the game will not have the desired results.

The game should also be fun. If it is enjoyable, trainees will want to take part and will be more involved in the activity, which will result in greater learning. Fun games will also keep their attention longer.

Practical Information for all Games

Before using any game in a safety training class, practice the game to make sure that 1) it works, 2) the game is not threatening or too difficult, 3) you have adequate time allotted, 4) the necessary materials are available and 5) you have the answers to any quizzes or other game where an answer is the result. You should also check that safety game instructions are clear and complete. If you need to stop to correct yourself or the directions, you will lose the trainees' attention and your credibility. For longer activities, clear written instructions might be necessary. Additionally, you should have more materials than you think you will need.

Copyright © 2006 SafetyFUNdamentals™

Props used in safety training games may get lost, break, or inevitably travel home with some employees. Basic materials include flip chart paper, pens, a laptop, and markers. Finally, make sure you will have enough room in the planned training space to have the trainees perform the activity. If they need to split up and work apart from each other, is there space in the room or will you have to send them elsewhere? If the teams need to spread out and work on large sheets of paper, is there enough table space? By working through the activity ahead of time, you will have an idea of the space required.

Dividing the training class into teams works well for most training games. Studies have shown that the social interaction that is involved with a team leads to greater learning. Teams also encourage more participation and ideas. The number of members on each team will affect the length of each activity. Teams of four or more often need a leader and usually take longer to work on an activity than a team of two or three.

Modifying Games and Activities to Make Them Yours

Most of the games and activities presented in this book can be modified to be used for a variety of topics. Once the basic structure of a game is established, it is usually only a matter of substituting the props and or instructions to make it applicable to a different topic in safety and health. You might also need to change the title of the game as well but often the trainer is the only one that sees the title. By using your imagination, you can often simplify, adapt, shorten or lengthen most games and activities based on the needs of the class and time constraints. When you do create a new game or activity or modify an existing one, make sure you incorporate all of the senses so that the activity appeals to all learners.

Copyright © 2006 SafetyFUNdamentals™

The Importance of Teams

You may have noticed that almost every one of these 77 games and activities involves putting the class members into small teams. Why is this? Team learning is part of accelerated learning which is explained in an earlier section of this chapter. Putting classes into teams is not only an easy way to incorporate aspects of accelerated learning into your classes but it can also make a significant difference in the quality of the learning that goes on. In 1983, Professors Donald Finkel and Steven Monk wrote about something called the "Atlas Complex." This idea refers to the thought that instructors feel like they are responsible for the entire learning process. By placing classes into smaller teams, the "Atlas Complex" can disappear and trainees can now be responsible for each other as well as themselves.

There are different ways that teams can be used in learning. In these 77 games and activities, most of the teams operate informally, i.e., little advance planning is required. When team activities are planned and used within existing course organization, the teams are considered more structured.

Informal teams can help break up "information dump," or when an instructor is talking for longer than 20 minutes straight. It can also be used to "wake up" trainees and get the trainees interested in the topic.

Structured or more formal team learning occurs when small groups are used very frequently. With structured teams, the way the group will be formed, the number of people in the group and the roles of individuals in the group are all carefully thought out in advance.

Teams vs. Groups

The idea of a team and the idea of a group are often used interchangeably. In this book, the word "team" is used for consistency although in some instances, what we are really talking about is a group. Any subset of people can be a group but only certain groups qualify as teams. While your training class may

Copyright © 2006 SafetyFUNdamentals™

initially be divided into smaller groups, these groups can become teams with your help.

Team members will have a high level of commitment to the group and a high level of trust within the group. For your groups to become teams, the trainees need time to interact with each other, resources (usually intellectual), a challenging task that becomes a common goal and frequent feedback on individual and group performance. Teams are able to work at a high level with individual team members also working at a high level. Team members can also challenge each other without causing problems and can be effective in accomplishing difficult tasks.

Team Size

Different experts suggest different team sizes but in almost all cases, the groups have less than 8 members. If the activity is very short, a smaller group, say 2 to 4 people, will work nicely. For longer more complex activities, a group of 5-7 may be necessary. It is also important to consider the size of the class. If there are 12 trainees, the activities would probably work better with 3 or 4 groups of 4 or 3, respectively, than 2 groups of 6 since there would be a definite "loser" and "winner" in games requiring competition.

Team Forming Ideas

Sometimes, creative methods are needed to put your trainees, who may or may not know each other well, into teams. If you divide the class by seating arrangement, you are likely to team people together who already know each other. While this may make them more comfortable, they may also be likely to act differently in the team than they would if they were teamed up with strangers.

You can divide the class by pure chance by using methods such as:

Candy/Gum Sort(1) - Place 4 packs of 5 different types of gum or candy into a bag. Pass the bag around the room asking each person to select one piece without looking. When they are finished,

Copyright © 2006 SafetyFUNdamentals™

you will have 5 random teams of 4. (Note: You can often get one large package of mixed candy or gum at a "Dollar Store")

Birthday Divide - Go around the room and ask everyone to shout out their birthday (month and day only). Write these on the board in chronological order. If you need teams of five, go down the list and put the first 5 into one team, the second 5 into a second team and so on.

You can also divide the class by chance but in a way that requires them to use some of the information they have learned in the class by using methods such as:

Matching Cards - Depending on the topic and number of trainees, write key words that go together on the back of index cards. Mix them up and place one at each seat in a classroom environment. When it is time to form a team, instruct the trainees to turn over their card and then find two or three other people (depending on how you set up the game) that have words associated with their word. For example, 4 key words that make up a team could be "confined", "PRCS", "space", and "rescue." The four trainees who each have one of these words would make up a team.

If this is a more formal class and trainees have name tags or place cards, you can hide the key word inside the place card or behind the name tag if it is in a plastic sleeve. This is better used after class material has been presented or reviewed so the trainees have familiarity with the words they are trying to match up.

MSDS Mates- For a slightly more complex sorting activity, you can use one MSDS for every group you need so if you need to form 4 teams, you will need 4 MSDSs. Separate all of the pages of the MSDSs (they should all have the same number of pages) and mix them together. The activity will be easier if the four MSDSs are for products that are fairly different from each other. Give each trainee one sheet of an MSDS. When it is time to form a group, tell the class that they need to put themselves in a group together with others who have a page of the same MSDS.

(1) All of these methods are based upon a class size of 20

Copyright © 2006 SafetyFUNdamentals™

Before putting the class into groups or teams, think about what you are trying to accomplish. If you want everyone involved, teams of three work best. If everyone must be involved, pairs are necessary. If the team is larger than 3, a leader will emerge and some may feel left out. Teams of four or five will also provide the opportunity for a leader to emerge. A team of 6 or more will require a definite leader with adequate leadership skills to lead the team. When assigning teams, keep in mind that some people are not comfortable being on the same team as their manager or supervisor and vice versa.

Copyright © 2006 SafetyFUNdamentals™

Prizes and Competition

Most of the activities in this book suggest giving a prize to the winning team. Usually these prizes are quite small - a candy bar would be fine. If your company is on a tight budget and you do not want to be springing for class prizes every week, there are a few very inexpensive or even free ways you can reward the winner. The following are only a few suggestions for you to consider. Use your imagination and you probably see prizes sitting all around your workplace.

- Paper certificates you can make on your computer and color printer
- Coupons for free coffee or tea in the company cafeteria
- A close parking space for a few days or week
- A sticker for a hard hat or cubicle or lunch box
- Permission to get in the lunch buffet line first (this works well for longer classes where the whole class will break at the same time to line up at a company supplied buffet line)

Most of the games and activities in this book involve a good amount of competition. Depending on your particular trainees, you might want to modify the activities slightly so that problems do not occur when very competitive employees are involved.

Copyright © 2006 SafetyFUNdamentals™

The Power of Laughter

Safety is serious business so what is there to laugh about? Plenty. Studies have shown that humor can have positive effects both physiologically as well as psychologically. Psychologically, humor can reduce stress and anxiety, enhance self-esteem, and increase self-motivation. What does all this mean? It is very difficult to train or teach a stressed out or anxious student. Additionally, it is very difficult if not impossible for the stressed out and/or anxious student to learn. Humor can help create a positive emotional and social environment which will help the trainee to lower his or her defenses and enable them to focus more clearly on the material being presented.

In 1986, William Glasser, M.D. proposed that the five primary needs of humans are survival, belonging, power, freedom AND fun. He also points out that all of our behavior is a constant attempt to satisfy one or more of those needs. In his book, "Choice Theory: A New Psychology of Choice Freedom," Glasser says that fun is difficult to define but it is associated with laughter, play and entertainment (which are the components of most of the 77 activities described in this book!).

> **Humor is also a way of saying something serious**
> **- T.S. Eliot**

Copyright © 2006 SafetyFUNdamentals™

Researchers have also shown that instructors who use humor are rated better by their peers and students. Wouldn't you like it if trainees really wanted to come to your classes? Trainers who use humor also show that they connect to the trainees which helps to form a bond. Other studies have shown that mangers get promoted faster if they have a good sense of humor.

> **A sense of humor is part of the art of leadership, of getting along with people, of getting things done.**
>
> **- Dwight D. Eisenhower**

Additional research also suggests that humor can increase student's interest in learning and that students taught with the use of humor will have better retention of the information as well. If the classroom is more relaxed and spontaneous humor is welcome and part of the class, trainees may be more wiling to participate and risk making mistakes as well since mistakes are more acceptable in this type of atmosphere.

Physiologically, humor and laughter affect the trainee by improving respiration and lowering their pulse and blood pressure. Laughter is also commonly associated with the release of endorphins into the bloodstream. Studies in the medical world have looked at how laughter can help patients heal by lessening stress and anxiety and increasing mental sharpness. Again, all great things to have in a training environment.

So, do you pop in a tape of Three Stooges and hope for the best or do you try to write your own stand up comedy routine? The answer, obviously, is neither. Humor can be a tricky subject. What one person might find funny, another might find offensive or stupid. Also, humor does also not equal telling a few jokes.

Copyright © 2006 SafetyFUNdamentals™

The best humor to use in your training classes will be specific, targeted and appropriate to the subject matter.

Humor is the great thing, the saving thing. The minute it crops up, all our irritations and resentments slip away and a sunny spirit takes their place.

- Mark Twain

Use appropriate cartoons and props. Do not use jokes that discriminate, are aggressive or put anyone down. Do not use anything that relates to sex, politics, illegal activity or any other organizational subjects that are taboo. When in doubt - throw it out!

To add some humor, consider starting classes with a related joke or comic. One great thing about a funny story or joke is that people who hear it often retell it which enhances the learning process (if it relates to the content). Also try to use jokes and stories that are practical. What is something you know that bothers your class? What do you all have in common?

Consider telling a funny story. Doug Stevenson, author of the "Eight Steps of Story Structure," says that the steps in creating a story include setting the scene, introducing characters and relationships, describing the goal and the obstacles to overcome and finally how the obstacle was overcome along with brining closure to any loose ends. He also says that you should make a point with a "phrase that pays" - something that will be remembered for some time. He also suggests asking if the situation in the story has ever happened to anyone in the class. This can help the trainees relate to the story and make the connection to the information you are presenting.

Copyright © 2006 SafetyFUNdamentals™

> **Storytelling is the most powerful way to put ideas into the world today.**
>
> **- Robert McAfee Brown**

If you are always training employees from your own facility that you know very well, chance are you know their style and what type of humor would work best. Select the kind of humor you use based on the personality and culture of your classes and company.

Safety is serious business. Although you take safety seriously and you encourage others to do the same, take yourself not so seriously and humor will fall into line.

Copyright © 2006 SafetyFUNdamentals™

Copyright © 2006 SafetyFUNdamentals™

Part 2

Games and Activities

Copyright © 2006 SafetyFUNdamentals™

Copyright © 2006 SafetyFUNdamentals™

How these 77 Games and Activities are Organized

Each of these 77 games and activities is explained over two pages. The first page is a summary page and provides the following information:

"Objectives"

Every good training class and every good training activity needs objectives. A generic objective for each activity is shown in this area but you can easily modify it depending on how you use the activity. You need an objective or two to determine if the activity was worthwhile. After every class where you use an activity, decide if the activity met your objectives.

"Good For"

The "good for" section contains suggestions for either the type of class or the subject matter where the activity would work well. This is by far not an all-inclusive list but merely a suggestion or two. You will probably find many other areas in which to use the activity.

"Class Length"

The "class length" section is again just a suggestion that is based on the length of the activity. If a class is only 30 minutes long, an activity that takes 20-30 minutes probably would not fit into your class as it is currently presented.

"Audience"

The "audience" section refers to the type of trainees that the activity is suited for but as you will see, most of these activities work well for all audiences.

Copyright © 2006 SafetyFUNdamentals™

"Where it Can Be Done"

Some of these activities need to be done indoors or in a classroom environment for practical reasons. It does not mean that an activity couldn't be done off-site, say at a construction site, but it may be a little more difficult in this environment.

"Team Size"

The "team size" shown in this section is a recommendation based on the logistics of the activity, the materials needed and the likely number of class attendees. You can probably use other team sizes if necessary but try the activity out first to make sure the timing is still correct. Team sizes are also based on recommended team sizes as described in the section on teams in Chapter 1.

"Time Needed"

The amount of "time needed" for an activity will vary based on a variety of factors including complexity, set-up, amount of instructions needed, background of the trainees, and the topic covered. The time needed is an estimate to get you started when planning your class. In most cases, the time needed can be shortened or lengthened by modifying the activity.

"Set-Up/Materials"

The "Set-Up/Materials" section is provided so that you can be well prepared to use these activities in your classes. You will see that in some cases, all that is needed is a writing instrument and paper or even a flip chart and markers. In other activities, more advance planning and preparation are needed.

Copyright © 2006 SafetyFUNdamentals™

The second page of each Activity description provides detailed **"Instructions"** and a description of the activity and any pre-class set-up that is required. In some instances, alternatives are suggested.

Suggested "**Debriefing/Discussion**" questions are also provided. It is sometimes tempting to skip this portion of the training activity but without it, all of your effort may have been for nothing. The debriefing is necessary to help the trainees relate what they learned or experienced during the exercise back to their workplace and day-to-day jobs. Feel free to add your own debriefing and discussion questions as well. The more ways you can tie the activity back to the trainees' actual job and responsibilities, the more effective the activity will be.

Copyright © 2006 SafetyFUNdamentals™

Copyright © 2006 SafetyFUNdamentals™

Let The Games Begin!

Copyright © 2006 SafetyFUNdamentals™

Copyright © 2006 SafetyFUNdamentals™

A Hammer, a Nail and a Screwdriver

Objectives	To encourage trainees to share challenges and provide suggestions
Good For	Large classes
Cautions	Do not let the teams play with real tools - it could be dangerous!
Class Length	At least 30 minutes
Audience	Any but reading and writing skills are necessary
Where it can be done	Anywhere
Team Size	Larger teams are better
Time Needed for activity	10 - 15 minutes
Set-Up/Materials	Paper, container and tools

Copyright © 2006 SafetyFUNdamentals™

Instructions for a Hammer, a Nail and a Screwdriver

Ask each team to think of one challenge they have that is associated with the topic being discussed and to write it on a slip of paper. Ask them to put these slips of paper into a box. Next, give each team a hammer, a nail and a screwdriver. If you don't have the real thing, you can use pictures of each taped to an index card or plastic toy versions. After each team has "tools", ask them to pick a slip of paper from the box (not picking their own). Tell the class that they will have 10 minutes to review the challenge and then use one or more of their tools to fix it. The tool can be used as intended or figuratively such as using a hammer to hammer out problems. After the 10 minutes, call stop and ask each team to share their challenge and their solutions.

Debriefing/Discussion Questions:

Which tools were most useful and why?

Did the use of these tools help you think in a different way?

Did having an outsider's perspective of your problem give you any new ideas?

Copyright © 2006 SafetyFUNdamentals™

And the Envelope Please

Objectives	To practice developing possible conclusions from accident information
Good For	Accident and Injury classes
Cautions	Reading required
Class Length	At least 30 minutes
Audience	Any
Where it can be done	Anywhere
Team Size	3 - 5
Time Needed	At least 15 minutes
Set-Up/Materials	Pre-written scenario cards

Copyright © 2006 SafetyFUNdamentals™

Instructions for And the Envelope Please

Before class, write out 10-12 different scenarios related to the topic. Put these scenarios on index cards. Put 5 or 6 of the scenario cards into an envelope. You will need an envelope for each team.

Divide the class into 2 groups. Give one "scenario" envelope to each team (each labeled appropriately). Have team members answer the questions on each scenario card. Allow 10 minutes for the exercise, then have each team share the scenarios with the class.

Note: You will need to develop about 5 or 6 separate scenarios for each envelope. An example of a scenario that could be used for a Bloodborne Pathogen training class would be:

> Upon entering the locker room, you find that Joe has apparently fallen and hit his head on a sink. He is lying unconscious on the floor and a small amount of blood is on his face and shirt. What should you do and why do you perform each step?

Debriefing/Discussion Questions:

How did reading a scenario help you to apply what you learned in this training class?

What additional safety precautions did you think of?

Was is it difficult to determine the correct answer? If so, why?

Copyright © 2006 SafetyFUNdamentals™

A-Z Safety Race

Objectives	To review information
Good For	Any topic
Cautions	Participants with poor English skills may have more difficulty. Note: you could allow them to complete the race using words from any language
Class Length	Any
Audience	Any
Where it can be done	Best in a classroom or in an environment where there are tables
Team Size	Any
Time Needed	5 - 10 minutes
Set-Up/Materials	Large pieces of paper (such as flip chart paper) , markers and a timer (flipchart standard optional)

Copyright © 2006 SafetyFUNdamentals™

Instructions for A-Z Safety Race

Divide the class up into teams of 3-6. Give each class a large sheet of paper and a supply of markers. Ask them to write the alphabet, A - Z, down the side of the page. If it is easier, they can split the page into two columns (A-M and N-Z). Tell the class that when you say go, they will need to write down a word related to the training topic that begins with each letter. The team that finishes first will win. Inform the teams that they are free to use their notes during the race.

Note: Don't hesitate to be lenient with the words teams submit. In some cases, this will even be necessary such as when trying to think of a word for the letter "X". Most teams will use "X-cellent," "X-tra," or other similar word.

After a team has finished, stop the other teams and review as a class. With a blank sheet of paper posted in front of the room, (preferably on a flip chart), call out each letter and ask the winning team to yell out their answer for each letter. After each answer is called, ask if there are any other answers that would also work. Write several of the answers given for each letter on the flip chart in the appropriate location. This will provide an even broader review of the material covered.

Debriefing/Discussion Questions:

How did this exercise help you to remember information in the class that you may have forgotten?

How did working in a group help you to finish faster?

What is one of the most important things you remembered as a result of this exercise?

Copyright © 2006 SafetyFUNdamentals™

Better Bingo

Objectives	To review material covered in class
Good For	After lunch or a break when trainees need an energizer
Cautions	Be careful not to go too fast. Basic reading is required
Class Length	Any but better for longer classes
Audience	Any
Where it can be done	Anywhere
Team Size	Individual Activity
Time Needed	Approximately 10-15 minutes for a standard game card
Set-Up/Materials	A Better BINGO card for each trainee and clues pre-written on small pieces of paper

Copyright © 2006 SafetyFUNdamentals™

Instructions for Better Bingo

Before class, design a BINGO card made up of 25 words and phrases related to the topic. Write out clues for each word or phrase and put on separate pieces of paper and then put these papers into a bag or box.

Give each trainee a Better BINGO card. (It is better if each card, (or at least most of them), is different so everyone does not win at the same time). Put the individual clues into a container so they can be selected randomly. Begin by picking a clue out of the container and reading it to the class. Tell the class that if the answer to the clue is on their Better BINGO card, they should mark it off. Tell the trainees that they should yell "BINGO" when they have crossed off all of their boxes in either a "T" or an "X" shape. Continue reading clues until someone yells BINGO. When there is a winner, you should ask the winner to read back the boxes they had checked off and then ask the rest of the class to help remember the clue that was read for that box. If possible, give the winner a small prize. See Appendix A for a sample BINGO card and clue sheet that can be used in an Office Ergonomics class.

Debriefing/Discussion Questions:

Which clues helped you to remember something that you may have forgotten?

How does hearing the clue and seeing the answer help you to learn?

Would this activity have been easier or harder if you worked in a team and why?

Copyright © 2006 SafetyFUNdamentals™

Brain Dump

Objectives	To review class material
Good For	Any subject
Cautions	Basic writing skills required
Class Length	Any
Audience	Any
Where it can be done	Anywhere
Team Size	2-4
Time Needed	Approximately 5-10 minutes
Set-Up/Materials	Paper and pencils/pens

Copyright © 2006 SafetyFUNdamentals™

Instructions for Brain Dump

Divide the class into teams of 2-4. Give each team a sheet of paper and tell them to number it 1 - 20. Tell them when you say go, their team will have 3 minutes to write down as many key phrases, relating to the topic, as they can think of (each phrase must be more than one word). If they can think of more than 20, they should keep adding to their list. After three minutes, call time and ask each team to report how many key phrases they were able to write down. Ask the team that reported the most key phrases to read off their list. If anyone in the class has any questions or doubts the legitimacy of any of the phrases, they should be encouraged to ask the winning team to explain.

Debriefing/Discussion Questions:

What was the most important thing you remembered after completing this exercise?

How did "Brain Drain" help you to remember some aspects of the training class that you may have forgotten?

What creative ways did you use to think of the phrases that you listed?

Copyright © 2006 SafetyFUNdamentals™

Camp Out

Objectives	To reinforce learning principles or to determine what information trainees need to know
Good For	Any topic
Cautions	Reading required
Class Length	Any
Audience	Any
Where it can be done	Classroom (table tops necessary)
Team Size	Pairs
Time Needed	10 minutes
Set-Up/Materials	Heavy cardboard-weight paper

Copyright © 2006 SafetyFUNdamentals™

Instructions for Camp Out

Give each trainee a "tent" (a piece of heavy paper or lightweight cardboard folded in half). If you are using this activity for an opener, particularly in a class where the topic is new to most trainees, tell the trainees that they should write one question (legibly) on one side of the tent that they would like answered during the class. Collect the tents and save them for the end of the class.

If the class is very familiar with the subject matter, such as with annual refresher classes, you can ask them to write 3 - 5 quiz questions on one side of the tent.

At the end of the class, redistribute the tents (trainees should not get their own card back). Each trainee should have one card with the question facing them. Each trainee should read the question in front of them to themselves and tell the answer (not the question) to their partner. The partner hearing the answer must try to guess the question.

Debriefing/Discussion Questions:

How did writing out what you needed to know help you focus during the class?

Was it difficult or easy to provide an answer instead of question to your partner and why?

What did you learn by reading someone else's question?

Copyright © 2006 SafetyFUNdamentals™

Can - Can

Objectives	Hands-on labeling practice
Good For	Practice proper labeling of containers
Cautions	When using cans, make sure the edges are not sharp or jagged or otherwise cover them to prevent an accident.
Class Length	At least 45 minutes
Audience	Any
Where it can be done	Anywhere
Team Size	2-4
Time Needed	Ten minutes
Set-Up/Materials	An empty clean can or similar container for each team , labels, pens, MSDSs, numbered list of company products or generic materials and timer

Copyright © 2006 SafetyFUNdamentals™

Instructions for Can - Can

Before class, write out a list of company products or generic chemicals. Divide trainees into pairs (or ask them to select a partner). Give each team a label to complete (hazardous waste labels, right-to-know labels, company labels, NFPA or HMIS labels can all be used). Going around the room, assign the first pair the first material on your list, the second pair the next item and so on. The trainees should imagine that material described on the card you gave them is inside their container. Tell the class they will have 10 minutes to correctly fill out their label. MSDSs and other documentation should be readily available.

To make this activity more competitive, you could write the materials on a flip chart for everyone to see and instead of assigning a material to each team, tell each team they can select the products they want to write a label for but since this is a contest, they should try to correctly write a label for as many products on the list that they can in the allotted time. The team with the most correctly completed labels at the end of the allotted time will be declared champion.

Debriefing/Discussion Questions:

How did trying to work quickly effect your ability to label the can correctly?

What was most useful resource in successfully completing a label?

What did you learn here that you can use back on the job?

Copyright © 2006 SafetyFUNdamentals™

Class Trip

Objectives	Practice for team safety inspections
Good For	Focused and general safety inspections
Cautions	Additional PPE may need to be made available
Class Length	At least 1 hour
Audience	Any
Where it can be done	Anywhere
Team Size	3-5
Time Needed	At least 30 minutes
Set-Up/Materials	Prepared inspection checklists

Copyright © 2006 SafetyFUNdamentals™

Instructions for Class Trip

Organize the class into teams. Present each team with a specific checklist they are to use when going on the "class trip." The checklists can be geared towards behavior, a specific area or a specific type of hazard. Instruct the team that they have 20-30 minutes to complete their trip and return to the class to present their findings to the class. Tell the class that they must also stay together as a team.

Debriefing/Discussion Questions:

How did working as a team differ from completing the exercise on your own?

What was the hardest thing about the class trip exercise?

How did using an inspection checklist help or hurt you?

Copyright © 2006 SafetyFUNdamentals™

Clockwork

Objectives	To think about special safety concerns associated with different times of day
Good For	Activities that occur around the clock
Cautions	None
Class Length	At least 45 minutes
Audience	Any
Where it can be done	Anywhere
Team Size	2-4
Time Needed	Approximately 15 minutes
Set-Up/Materials	"Clock sheets" for each team

Copyright © 2006 SafetyFUNdamentals™

Instructions for Clockwork

Divide the class into teams of 2-4. Give each sheet a blank "Clock Diagram." A sample blank Clock Diagram can be found in Appendix B.

In a general safety class, ask each team to think of specific hazards related to each time of day. They should write as many as they can think of inside the appropriate area of the clock diagram. If the class topic was related to a specific process or an activity that has very specific steps over the course of the day, the teams should be asked to write down the hazards that may be present during each quarter on the clock. After 10-15 minutes, ask each team to share their results.

Debriefing/Discussion Questions:

Why may different hazards exist at different times of the day?

How did working with the clock diagram help you to think about these hazards differently?

Which hazards exist equally over all of the quarters?

Copyright © 2006 SafetyFUNdamentals™

Confidence

Objectives	Team review of class material
Good For	A review of any topic
Cautions	Some individuals may become too competitive
Class Length	At least 45 minutes
Audience	Any
Where it can be done	Classroom
Team Size	Any
Time Needed	25 minutes
Set-Up/Materials	Sets of prepared questions

Copyright © 2006 SafetyFUNdamentals™

Instructions for Confidence

This is a competitive team game where teams predict how well they learned the content of the training class. Before the class, write out 30 test questions based on the material you covered (or will cover). Divide the questions up into groups of ten and put each group of ten on a separate piece of paper. Make a copy of each set of questions for each team. Don't forget to develop an answer key for each set of questions as well. Tell each team that they will be given 5 minutes to answer the 10 questions on the first sheet. Before giving them the first set of questions, ask them to estimate how many they think they will answer right and post that number on a flip chart or board. Distribute the first set of questions and tell the teams to start. After 5 minutes, tell the teams to stop and review the correct answers. Ask the teams to calculate their scores by giving ten points to each correct answer. Play a second and third round just like the first. At the end of the three sets of quizzes, calculate the point totals and declare two winners - one who estimated their correct scores the best and one with the best score.

Debriefing/Discussion Questions:

How did trying to predict your score make you answer questions differently?

How did it affect teamwork?

Why did you like or not like being put on the spot and asked to talk about how confident you were of your answers?

Copyright © 2006 SafetyFUNdamentals™

Crazy Crossword

Objectives	To review information
Good For	Any topic
Cautions	Reading and writing required
Class Length	At least 30 minutes
Audience	Anyone familiar with the content
Where it can be done	Classroom
Team Size	2 - 4
Time Needed	Varies depending on difficulty of content
Set-Up/Materials	Completed crossword puzzle

Copyright © 2006 SafetyFUNdamentals™

Instructions for Crazy Crossword

Divide the class into teams of 2-4. Give each team a "crazy crossword." A crazy crossword looks like a completed crossword puzzle but in the area where there should be clues for across and down, there are only numbers. Present the across and down section on a separate page. If possible, have several different versions of the puzzle, especially if you plan to make this activity more complex by following the optional steps below. Tell each team they have 10 minutes (this can be longer depending on the complexity of the words) to write down appropriate clues for each of the words (they are basically doing the puzzle in reverse).

If time allows and you want to take this activity one step further, follow the optional steps below.

When all "clues" are completed by each team, collect the clue sheets and mix them up. Now give each team a copy of another team's clues along with a blank version of the puzzle that matches those clues. Tell them they can race to see who can complete the new puzzle the fastest.

Note: If you need blank crossword puzzle templates to work with, you can download them at www.crownsafety.com. (See instructions on page197)

Debriefing/Discussion Questions:

Why was it easy or difficult to do the puzzle in reverse?

How did working as a team help you?

If there were any words or phrases you did not know, what were they?

Copyright © 2006 SafetyFUNdamentals™

Cryptogram

REKBMQZETL TRMXDXMK

Key: K = Y

Objectives	Warm up or welcome exercise
Good For	Introducing a topic and encouraging teamwork
Cautions	Can be difficult for many people
Class Length	Any
Audience	Literacy is important for this exercise
Where it can be done	Anywhere but classroom is best
Team Size	1-3
Time Needed	At least ten minutes
Set-Up/Materials	Cryptogram (with at least one letter hint) for each trainee or team

Copyright © 2006 SafetyFUNdamentals™

Instructions for Cryptogram

Puzzles like cryptograms and crossword puzzles can be a great way to start a refresher course. A puzzle such as a cryptogram can be placed at each person's spot before they enter the room with instructions to begin working on the puzzle when they sit down. If they finish, they are to help someone else with their puzzle. A sample cryptogram from a Housekeeping class can be found in Appendix D.

Debriefing/Discussion Questions:

Why do you like or dislike cryptograms?

How did the cryptogram get you to begin focusing on the topic?

How did working with a classmate help you to complete the puzzle?

Copyright © 2006 SafetyFUNdamentals™

safetyFUNdamentals

Domino Effect

Objectives	To demonstrate how one act influences future acts and results
Good For	Training where a series of events are involved such as process safety
Cautions	None
Class Length	Any
Audience	Anyone
Where it can be done	Anywhere but a flat surface is needed
Team Size	2-4
Time Needed	15-20 minutes
Set-Up/Materials	Dominoes and white labels cut in half

Copyright © 2006 SafetyFUNdamentals™

Instructions for Domino Effect

Divide the class into teams of 2-4 trainees. Give each team a set of dominoes (or at least a dozen) and a supply of small white labels. Tell each team that they need to break down a process being discussed in the class into small steps and then write each of these steps onto a label. Each label should be placed on a separate domino. Depending on the class topic, you may need or want to assign a process to each group. (It can be the same process for each group.) After approximately 20 minutes (the necessary time will depend on the complexity of the processes) ask each team to line up their dominoes vertically and in the correct order. When you say go, they should tip over their first domino and watch the chain reaction. Ask each group to now stand up their dominoes again and remove one or two without moving the remaining dominoes any closer to each other. Ask them to start the chain reaction again and note the differences. Ask them to try removing various dominoes in key positions, such as the first and the last, and also see what happens.

If each team worked on a different process, you can now go around to each team and ask them to describe the steps in their process.

Debriefing/Discussion Questions:

Why was it easy or difficult to break the process down into steps?

What effect did taking out a few of the steps (dominoes) have on the process? Would this be the same result in the real world?

Did removing either the first or last step make a big difference? Why?

At work, what else could change the flow of steps?

Copyright © 2006 SafetyFUNdamentals™

Draw It Out

Objectives	To visually review important words or concepts from the class material
Good For	Any topic
Cautions	None
Class Length	Any
Audience	Any
Where it can be done	Anywhere a drawing can be taped to a wall
Team Size	2-4
Time Needed	Approximately 15 minutes
Set-Up/Materials	Paper and markers for trainees, tape

Copyright © 2006 SafetyFUNdamentals™

Instructions for Draw it Out

Divide the class attendees into teams of 2-4. Assign each group 2-3 terms that relate to the class topic and ask them to draw a picture of that term on a sheet of paper. There should be one drawing per page.

After 5 minutes, either tape all of the drawings on the wall or pass them around. (In a small group you can hold them up). Ask the other teams to guess and explain what each picture represents.

For example, a "*Words to Draw*" list for a class on machine guarding might include:

- ✓ Pinch Point
- ✓ Light Curtain
- ✓ Feeder Stick
- ✓ Guard
- ✓ Tool Rest

Debriefing/Discussion Questions:

How did drawing help you to think about safety issues related to the topic?

When drawing, what item did you have trouble remembering?

Why is it easier or harder to convey a safety message by drawing than by just explaining it to someone?

Copyright © 2006 SafetyFUNdamentals™

Extra! Extra! Read All About It!

Objectives	To use the information presented in class in predicting positive and negative outcomes related to the content
Good For	Any topic
Cautions	Good writing skills required
Class Length	Any
Audience	Any
Where it can be done	Classroom
Team Size	2-4
Time Needed	Approximately 20 minutes
Set-Up/Materials	Paper and markers for each team, tape

Copyright © 2006 SafetyFUNdamentals™

Instructions for Extra! Extra! Read All About It!

After putting the trainees into teams of 2-4, tell them their assignment is to come up with two newspaper headlines related to the class. One headline is to be negative, that is, for something that went very wrong and the second headline should be something positive that happened. When the headlines are finished, they should be written on large sheets of paper and posted. Review the headlines as a class.

Option:
Teams could switch headlines and work on developing a short story to go along with each headline. After 10 minutes, the stories should be posted and reviewed by the class as a whole.

Debriefing/Discussion Questions:

What type of actions would have had to have taken place (or not) for the headline to occur?

How could a headline mislead you on the cause of the accident?

Where did you get your ideas for the bad headlines?

How did reading possible bad headlines help you to understand possible things that could go wrong?

Copyright © 2006 SafetyFUNdamentals™

safetyFUNdamentals

Flashback

Objectives	To demonstrate how different people see things differently and how this can influence safety inspections
Good For	Safety inspections and observation training
Cautions	None - if writing skills are an issue, only one scribe per group is necessary
Class Length	Any
Audience	Any
Where it can be done	Anywhere
Team Size	4-6
Time Needed	15 minutes
Set-Up/Materials	A large (at least 8" x 10") photograph of a work area

Copyright © 2006 SafetyFUNdamentals™

Instructions for Flashback

Divide the class into teams of 4-6 and assign one team member in each group to be the "flasher." Have each team gather around an area of the table. Give each tea m the same prepared enlarged photograph of an area in the workplace and place it face down on the table in the middle of the group. Give each team a piece of paper and pen or pencil as well. Tell the team that they will have 30 seconds to study the photo as best they can. The "flasher" will turn the photo over when you say go. After 30 seconds, you will say "stop" or ring a bell and the "flasher" will turn the photo facedown once again. The teams are now to work together to list as many things as they can that were in the photo that relate to safety. (You can specify what type of items should be on the list depending on the type of class you are delivering and the type of photos you have access to). After several minutes, which should be more than enough time, tell the teams to stop making their list. Go around the room and ask each "flasher" to report the number of items they found. The team with the most number of items wins. If you had instructed the team to list general items, such as anything related to safety, you should also ask the winning team to tell why the item they listed relates to safety.

Debriefing/Discussion Questions:

Why did some teams see different things than other teams?

What was the hardest thing about this exercise?

How do you think this applies to safety observations in the workplace?

Copyright © 2006 SafetyFUNdamentals™

Flying Boulders

Objectives	To get the participants interested in the topic initially and to start their minds thinking about the information to be covered
Good For	Any topic but especially those that have been covered before
Cautions	"Boulders" should not be thrown at close range, at another's face or forcefully!
Class Length	Any
Audience	Any
Where it can be done	Classroom settings work best
Team Size	At least 10 trainees
Time Needed	About 1 minute per trainee
Set-Up/Materials	White or gray paper and pencils/pens

Copyright © 2006 SafetyFUNdamentals™

Instructions for Flying Boulders

Give each trainee a piece of 8.5" x 11"paper. Ask the trainees to write a question on their paper that relates to the topic. Next, ask the trainees to scrunch their papers into a ball (or "boulder"). Divide the class into two teams and ask them to face each other. When you say start, tell the trainees to throw their boulders to the other team. If they catch another boulder or one falls near them, they are to pick it up and throw it back to the other side. Let this "boulder toss" continue for about 1 minute before calling stop. Each person is to sit down again and open their "boulder". As time allows, go around the room and ask each person to un-crumple their paper and read the question. If they can answer their question, they should. If not, anyone in the class should answer.

The questions that are asked and the answers that are provided will give you a good idea of what information needs to be covered or focused on.

If questions were written on each "boulder", you can ask each trainee to make it his/her job to find the answer to that question during the course of the class. At the end, each person can reveal their question and answer as a short review activity.

Debriefing/Discussion Questions:

Are you surprised at how much you remembered from a previous class?

What information did the exercise remind you of that you had forgotten?

How did being responsible for learning and reporting on one piece of information affect your class participation?

Copyright © 2006 SafetyFUNdamentals™

safetyFUNdamentals

Forklift Frenzy

Objectives	To review forklift safety information
Good For	Material handling topics
Cautions	A lot of preparation for the trainer
Class Length	At least 45 minutes
Audience	Any
Where it can be done	Classroom
Team Size	3-4
Time Needed	Approximately 20 - 30 minutes
Set-Up/Materials	Forklift Frenzy game track (see Appendix), prepared question cards , mini-forklift for each team or player (if mini-forklifts aren't available, any type of game piece will do)

Copyright © 2006 SafetyFUNdamentals™

Instructions for Forklift Frenzy

Before class, write out at least 20 questions on 20 index cards. Each card should also have a number on it from 1 to 6. Four sheets of flipchart paper should be taped together end to end so that one long "forklift track" is running down the middle of the training floor. The 20 or so cards should be shuffled first and placed face down in a line starting at one end of the track and running to the other. A sample set up is shown in Appendix F. All teams should start with their forklift at the start position. The first team turns over the card closet to them. If they answer (as a team) the question correctly, they then get to move the number of spaces shown on the card. They should move that many spaces and land on the appropriate card (but NOT turn it over). The second team now turns over the next card closest to the start line and does the same. If they answer the question right, they also get to move the number of spaces shown. If the number of spaces shown requires them to land on the same card as the first team, the first team gets knocked off and sent back to the starting line. If not, the second team gets to stay on their card as well without turning it over. The 3rd and fourth teams (if there are 3rd and 4th teams) take their turns in the same fashion. After all four teams have had their first turns, the first team either turns over the card they had been be standing on and answers and moves again as in the first round or, if they had been bumped back to the starting line, they turn over the next closest card and continue as in round one. When one team reaches the finish line, the game is over and the winning team announced.

Debriefing/Discussion Questions:

What questions helped you to remember something about forklift safety that you may have forgotten?

How did your team work together to decide on the correct answer?

What additional safety information should have been included?

Copyright © 2006 SafetyFUNdamentals™

Four Square

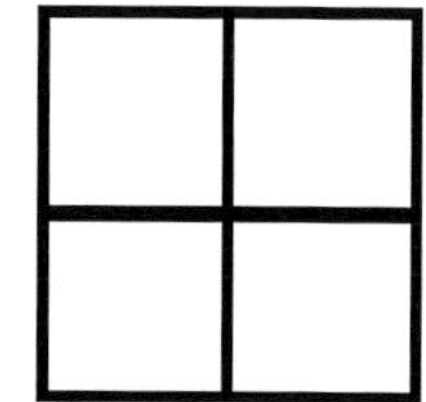

Objectives	To review material in a chemical safety class
Good For	Chemical safety topics
Cautions	This is a more advanced game than most.
Class Length	At least one hour
Audience	Good reading skills are required
Where it can be done	Classroom
Team Size	2-4
Time Needed	Approximately 30 minutes
Set-Up/Materials	Game board for each team, one dice for each team, one playing piece, MSDSs.

Copyright © 2006 SafetyFUNdamentals™

Instructions for Four Square

Divide the class into teams of four. Give each team a game board, a dice and a supply of MSDSs - at least enough for each player to have several. A sample Four Square board can be found in Appendix E. Each of the four players should have the same supply of MSDSs. Each player should start with their marker on a square of the board and the youngest player should go first. The first player rolls the dice and moves their marker that many spaces around the board. Another player must ask the first player a question taken from one of the MSDSs relating to that square (for example, if the first player landed on "health", a health-related question must be asked). The "questioner" should tell the player which MSDS he or she is getting the question from. The player can use his or her copy of the MSDS to answer the question. If the question is answered right, the player should write his or her initials in that box. The winner will be the first player to place his or her initials in all four squares. After the first player has answered the question (either right or wrong), the player to the left takes a turn and the process is repeated. The games continues until one player has answered a question correctly in each of the four categories. To encourage collaboration, each game can be played by a team of two playing against a second team of two instead of four individuals playing against each other.

Note: The four areas on this board are 1) Health, 2) Fire, 3) Reactivity and 4) Special/Miscellaneous. These categories can be changed to better fit in with your particular class

Debriefing/Discussion Questions:

What information did you learn that you can use immediately back on the job?

What was the most difficult thing about this exercise?

Copyright © 2006 SafetyFUNdamentals™

From Here To There

Objectives	To review steps in a process
Good For	Subjects and processes with steps or parts
Cautions	None
Class Length	At least 45 minutes
Audience	Any
Where it can be done	Classroom
Team Size	2-4
Time Needed	Approximately 10-15 minutes
Set-Up/Materials	Scissors, large paper and markers for each team

Copyright © 2006 SafetyFUNdamentals™

Instructions for From Here to There

When training a class on a particular topic that involves many steps, like completing an OSHA injury log, break up the topic into distinct steps. Before starting to teach the topic, give each team a set of scissors, a large piece of paper and markers. Assign each team a step of the process. Tell them they are responsible for capturing their step in an easily understood drawing (without using words) that will be used to give directions to anyone needing to correctly complete the steps in order. At the end of the lesson, give the class a few minutes to finish up and then ask them to post their step on the wall in the correct order. After all teams have finished, ask the rest of the class to describe the other team's steps as best they can.

Debriefing/Discussion Questions:

How did having to use pictures only effect your ability to describe the process?

Why do you think it is easier or harder to understand safety procedures by using pictures only?

How can you use this information back on the job?

Copyright © 2006 SafetyFUNdamentals™

Hand Bone Connected To The Wrist Bone

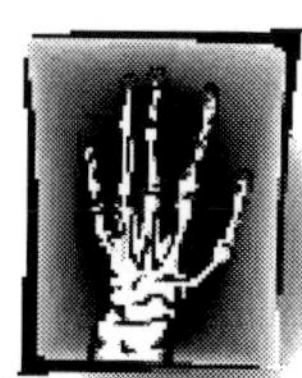

Objectives	To review body movements and related information
Good For	Wrist, hand and arm injury prevention training
Cautions	None
Class Length	At least 30 minutes
Audience	Any
Where it can be done	Classroom
Team Size	Any
Time Needed	20 minutes
Set-Up/Materials	X-rays (real or purchased) or large blown up photos or drawing or models of the hand and wrist

Copyright © 2006 SafetyFUNdamentals™

Instructions for Hand Bone Connected to the Wrist Bone

Divide the class into teams and give each team an assembly packet. Each packet should be made up of a model, either paper or an actual 3-D model that has been dismantled or cut into smaller parts. When you say go, ask each team to assemble their model as quickly as possible and tell them when they are through, they need to mark a particular area of possible injury that results from a particular activity affecting the hand and/or wrist. The area and the activity should be based on your specific workplace. For example, for a team of word processors, they could be asked to identify what part of the model could be affected by typing all day every day and be ready to describe that injury to the group.

Debriefing/Discussion Questions:

How does assembling the model help you to better understand how injuries occur?

Why was it easy or difficult to put the hand (or other parts) together?

What can you do differently to avoid this type of injury in the workplace?

Copyright © 2006 SafetyFUNdamentals™

Hazard Hunt

Objectives	To practice hazard recognition
Good For	Hazard recognition or safety inspection training
Cautions	Be careful that you are not setting up real hazards
Class Length	At least 15 minutes
Audience	Any
Where it can be done	Anywhere
Team Size	Any
Time Needed	15 - 30 minutes
Set-Up/Materials	A prepared work area or photo of a work area that shows a variety of hazards

Copyright © 2006 SafetyFUNdamentals™

Instructions for Hazard Hunt

Set up an area in your location that contains a number of **fake** hazards such as an unlabeled container (with water inside), tripping hazards, and other situations common to your workplace. If you want to get really graphic, you can add some fake blood (sold in gag shops). *Make sure your fake area is cordoned off from the rest of the workforce and that everything is put back to a safe condition immediately after the exercise.* If you cannot set up a fake unsafe work area for the class, consider setting up an area, take a good high resolution photo of it, and then provide the trainees with the photo instead. If this is also not possible, use a photo of any area that shows several hazards.

After putting the trainees into teams, tell them that they have 5 minutes to try to find as many hazards in the area of the photo as possible. After the 5 minutes, go around to each team and ask them to state one of the hazards they found. Compare and contrast the different hazards reported by each team.

Debriefing/Discussion Questions:

How did it feel to have a limited amount of time to observe hazards?

What was your first reaction when you saw so many hazards together?

How did working with a team affect your ability to identify hazards?

Which of the hazards you found would be a top priority to fix?

Copyright © 2006 SafetyFUNdamentals™

HazCom Hill

Objectives	To review OSHA's Hazard Communication Standard
Good For	Hazard Communication and MSDS review
Cautions	Reading required
Class Length	At least an hour
Audience	Any
Where it can be done	Classroom
Team Size	2-3
Time Needed	Approximately 30 minutes
Set-Up/Materials	HazCom Hill Game Board, supply of MSDSs, dice

Copyright © 2006 SafetyFUNdamentals™

Instructions for HazCom Hill

On a large piece of poster board, copy the HazCom Hill game board. A sample HazCom Hill game board can be found in Appendix G. Place the game board in the center of a table with all trainees gathered around it. For large classes, i.e. 20 trainees, you may want to make several game boards and have several games running at the same time.

Each team should have two 16-section MSDSs. One is the main MSDS they will play with and the other is the alternate. Each team should take turns rolling the dice to determine how many spaces to move. When they land on a square, the other team must ask them a question from that numbered section of the main MSDS and they must try to answer it. (When they get to the second half of the board, the squares have two numbers so they will need to answer two questions.) If they already answered a similar question, they should work from the alternate MSDS. Each team takes a turn and if they can successfully answer the question(s), they can stay on that square. If they cannot, they must return to where they started. The first team to make it "up the hill" (to the end) wins.

Debriefing/Discussion Questions:

What was the most difficult part of this game?

What is one thing new that you learned?

What new information can you use back in the workplace?

What do you think is the most difficult section of an MSDS and why?

Copyright © 2006 SafetyFUNdamentals™

Health Homerun

Objectives	To review health-related training information
Good For	Any health-related topic
Cautions	Some reading required
Class Length	Any
Audience	Anyone
Where it can be done	Anywhere
Team Size	Any
Time Needed	15-20 minutes
Set-Up/Materials	Prepared question cards, blank pieces of white paper for "bases"

Copyright © 2006 SafetyFUNdamentals™

Instructions for Health Homerun

Arrange the training room into a baseball diamond. You will need to identify home plate, first base, second base, and third base. Each team should select one player to be their "runner" and the remaining members of the team will be their coaches standing on the sidelines.

Place prepared question cards into a box. Twenty cards is a good number to start with but you can add more if you would like the game to be longer. You should also include a few cards that just say "strike" and a few that say "ball." If a player selects a "strike" card, he or she loses a turn. If the player selects a "ball" card, he or she gets to advance to the next base without answering a question. With every card drawn, the "coaches" get to help the runner to decide upon the best answer to the question.

The first player starting should select a card from the box and read it. If the card says strike or ball, he or she should move appropriately. (If a strike is drawn, the player loses a turn. If a ball is drawn, the player advances one base). If there is a question, the card should be handed to the umpire (the trainer/facilitator) who will read the question out loud. The runner can confer with his or her coaches to come up with the best answer. If the question is answered correctly, he gets to advance to the next base position. The second team now gets to go. The teams continue taking turns until all of the cards in the box have been selected and acted on. The team with the most "runs" (or complete circuits around the bases) wins.

Debriefing/Discussion Questions:

Why do you feel that some questions are worth more than others?

How did having a team of coaches help to get answers right?

Copyright © 2006 SafetyFUNdamentals™

safetyFUNdamentals

Help Wanted

Objectives	To review individual responsibilities for safety
Good For	Safety awareness and leadership training
Cautions	Writing skills required
Class Length	Any
Audience	Any
Where it can be done	Anywhere
Team Size	2-4
Time Needed	Approximately 10-15 minutes
Set-Up/Materials	Paper and pencils, sample "Help Wanted" Ads

Copyright © 2006 SafetyFUNdamentals™

Instructions for Help Wanted

Divide the class into teams of 2-4. Tell the class that they have the responsibility to write a "Help Wanted" advertisement for a new employee to work in their department. They need to find someone with the best safety experience for that operation. Their job is to write the advertisement so that the best (and safest) person gets the job. Tell them to write the advertisement being sure to list out the safety skills and experiences that are required. After every team is finished, go around the room and ask them to read their advertisement.

Note: You might want to have several old newspapers with classified advertisements to give them an idea on how to get started.

Debriefing/Discussion Questions:

Would it be difficult or easy to find someone to fill the position you described? Why?

How did your own experience help in writing the advertisement?

What is the most important characteristic that would make this person successful?

Copyright © 2006 SafetyFUNdamentals™

safetyFUNdamentals

I'm Puzzled

Objectives	To demonstrate the interconnectedness of different parts, people or departments
Good For	Complex topics that have many inputs
Cautions	Writing skills required; slightly more difficult than other activities.
Class Length	At least one hour
Audience	Any but reading and writing skills are necessary
Where it can be done	Classroom
Team Size	3-5
Time Needed	20 - 30 minutes
Set-Up/Materials	Purchased blank puzzle or one made out of poster board, pencils, markers, tape

Copyright © 2006 SafetyFUNdamentals™

Instructions for I'm Puzzled

Give each team a blank puzzle. Depending on the length of the class and the topic, puzzles should have 10 - 30 pieces. The team needs to design their puzzle in a way that all components of the topic are included and connect to the other pieces in the right order. Most puzzles are square but puzzles can also be linear if the process is better suited to a linear puzzle. Each team can be assigned a different process or the same process. After the teams have had some time to complete their puzzles, ask them to take the pieces apart and place them into an envelope with the topic written on the front of the envelope. Envelopes should be switched between teams and teams should then be asked to race to assemble the puzzles they were given (all puzzles should have the same number of pieces). After the puzzles are completed, the winning puzzle should be shared with the class. The winning team should receive a prize.

Debriefing/Discussion Questions:

Why did you find this activity easy or difficult?

Why did you feel that a puzzle was or was not able to represent the process well?

What new information can you use back in the workplace?

Copyright © 2006 SafetyFUNdamentals™

It's In The Bag

Objectives	To warm up the class or review items that relate to the class topic
Good For	Any topic
Cautions	Use only safe objects
Class Length	Any
Audience	Any
Where it can be done	Anywhere
Team Size	2-4
Time Needed	Approximately 10-15 minutes
Set-Up/Materials	Brown bag and various small safety-related items

Copyright © 2006 SafetyFUNdamentals™

Instructions for It's In the Bag

Prior to class, collect various small items (but nothing pointy, sharp or otherwise dangerous) and place them into a bag. You will need one bag for each team. Divide the class into small teams of 2-4. Give each team a prepared bag and tell them they should pass the bag around and without looking, feel each object in the bag and write down the name of each object they feel. The entire team should go through this process and should all agree on the final list. Remind the class that there is no peeking! If possible, each team should receive different objects in their bag but if this is not possible due to the subject matter, then at least a few of the bags should be different. After all teams are finished, ask each team to read their list of "found" items to the class and after they do so, they should dump out their bags and see if they were correct.

For a general safety class, the following items could be collected and put into a bag: a lock, a tag, a paper dust mask, a vinyl love, earplugs, a packet of thirst quencher, a Band-Aid, and anything else relatively small, unbreakable and otherwise not dangerous, that relates to the class topic.

Debriefing/Discussion Questions:

How did this activity get you to start thinking about the topic or act as a review?

If there was an item that you could not guess or guessed incorrectly, what was it and why do you think it couldn't be identified?

What would have helped you to identify all of the objects?

Copyright © 2006 SafetyFUNdamentals™

Jargon Jar

Objectives	Review of any topic
Good For	Any topic that contains a lot of jargon or acronyms
Cautions	Reading required. Make sure everyone knows what "jargon" means.
Class Length	Any
Audience	Any
Where it can be done	Anywhere
Team Size	1-3
Time Needed	10 minutes or more
Set-Up/Materials	Jargon that pertains to the class topic written on small slips of paper and placed inside a jar (or box, hard hat, or other container)

Copyright © 2006 SafetyFUNdamentals™

Instructions for Jargon Jar

Before the class, write out different words of jargon and acronyms that pertain to the topic on small slips of paper and place in a jar, hard hat or other container. Also keep one master list of the words that you used and their meaning. When you are ready to conduct the activity, explain to the class that different words pertaining to the topic are on pieces of paper in the container and you will be passing the container around. They need to pick a slip of paper to get their assignment. Tell the trainees that they will have 3 minutes to come up with a definition for the letters on their paper PLUS to come up with as many other possibilities for the jargon or acronym as they can within the three minute time frame. The trainee who thinks of the most possibilities for their word will win a prize. This game is meant to get them laughing while providing a review of the material covered. At the end of the three minutes, ask everyone to shout out how many "alternative" words they came up with. When a winner is found, ask him or her to share their words. When all teams are done, review all of the other words that were in the jar and their true meaning.

Debriefing/Discussion Questions:

Why do we use jargon in safety and health?

Which words did you learn today that you didn't know before?

When should jargon not be used?

Copyright © 2006 SafetyFUNdamentals™

Jars And Jellies

Objectives	To encourage teams to discuss and share information
Good For	Topics where you have information presented in categories, such as Chemical Safety
Cautions	The class may eat your supplies!
Class Length	Any
Audience	Any
Where it can be done	Anywhere
Team Size	3-5
Time Needed	Approximately 10 minutes
Set-Up/Materials	A large jar and jellybeans

Copyright © 2006 SafetyFUNdamentals™

Instructions for Jars and Jellies

Fill a large jar (such as an empty and cleaned mayonnaise jar) with jellybeans. If possible, only include one color jelly bean for each of the categories you covered in class. For example, if you are discussing chemical safety, you could have one color represent corrosives, one color represent acids, one color represent reactives, etc.

Divide the class into teams of 3-5. Pass the jar around and tell each trainee to take what they want and then to keep passing the jar around. (Do not tell them ahead of time how colors are related to categories). After each trainee has taken jellybeans, tell them to put all of their team's jelly beans into a pile. Tell the teams that their assignment is to think of a statement or fact for each of their color-coded jelly beans. When all teams are finished, go around to each group and ask them to share their facts.

Debriefing/Discussion Questions:

Was it easy or difficult to think of a fact for each one of your categories?

What was the most important fact you learned?

Copyright © 2006 SafetyFUNdamentals™

Lessons Learned

Objectives	To encourage trainees to make their own connections from information they hear to their work activities
Good For	Classes where trainees are very familiar with the content such as in annual refreshers
Cautions	Writing required
Class Length	At least 30 minutes
Audience	Any
Where it can be done	Anywhere
Team Size	2-4
Time Needed	Approximately 10 minutes
Set-Up/Materials	Prepared stories or case studies

Copyright © 2006 SafetyFUNdamentals™

Instructions for Lessons Learned

Tell the class a story that has the learning concepts you need to convey imbedded in the material (you can tell stories of your own or that you heard from others as long as they relate to the topic you are teaching). The story should be several minutes long. Tell the trainees that they should listen carefully to the story because it will not be repeated. You can also encourage the trainees to take notes while you are reading. After the story is completed, divide the class into teams of 2-4 and instruct each team to write down as many "lessons learned" from the story as they can. After 5 minutes, ask for a team to volunteer to share their lessons learned. When they are finished, ask if anyone else has any other lessons to share.

Debriefing/Discussion Questions:

How did you listen differently knowing that you would have to remember the information?

What was your favorite "lesson learned" and why?

What (if any) "lesson learned" did you forget to include on your list and why do you think this one was left out?

Copyright © 2006 SafetyFUNdamentals™

Lone Ranger

Objectives	To learn differences in perception of hazards
Good For	Housekeeping and general safety classes
Cautions	Some reading required
Class Length	At least 45 minutes
Audience	Any
Where it can be done	Anywhere
Team Size	2-5
Time Needed	Approximately 10 minutes
Set-Up/Materials	Prepared hazard cards (one set for each team)

Copyright © 2006 SafetyFUNdamentals™

Instructions for Lone Ranger

Divide the class into teams of 2-5. Give each team a prepared envelope containing 10 hazard cards. (Each team should receive the same cards in their envelopes). Ask each team to sort the cards into the order of importance as if there was only one individual available in the company who could repair everything. Which hazard would they make top priority, second priority, and so on? After each group finishes sorting their cards in priority order and makes their lists, the card piles should be compared. All differences should be discussed as a class.

For example, hazards that could be listed on cards separately could be 1) broken light bulb, 2) door hanging off hinges, 3) broken guard rail, 4) frayed electrical cord, 5) missing screws in equipment access panel, 6) loose and torn rug in office, 7) broken handrail on stairs, 8) missing regulator from air hose, 9) non-functioning alarm speaker, and 10) broken ladder rung.

Debriefing/Discussion Questions:

Why do you think teams could not agree on what condition is most hazardous?

What would make it easier for individuals and teams to be able to prioritize hazards in the same order? Is this a good thing?

If there was only one person making repairs, how could he or she best prioritize what to do?

Copyright © 2006 SafetyFUNdamentals™

Match Maker, Match Maker

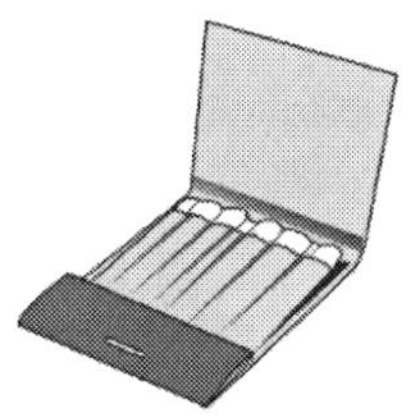

Objectives	To demonstrate understanding of proper tools and equipment for particular tasks
Good For	PPE training; hand tool safety
Cautions	None
Class Length	At least 45 minutes
Audience	Any
Where it can be done	Anywhere
Team Size	2 teams
Time Needed	Fifteen minutes
Set-Up/Materials	Various types of PPE, tools and products

Copyright © 2006 SafetyFUNdamentals™

Instructions for Match Maker Match Maker

Before the class, gather samples of different kinds of PPE and tools such as:

2 sets of Rubber Gloves; 2 sets of Latex Gloves; 2 sets of Canvas Gloves; 2 sets of gloves made of other material; Grease-based product; Corrosive product; Alcohol-based product; Object with sharp edges or one that may splinter; Other products as available.

Place the products on a table in the front of the room. Divide the class into two teams and give each team one each of the various types of gloves. Tell the group that each team should place the appropriate glove for each product in front of it and whichever team gets all products right first wins. Tell the class that some products might share the same type of glove.

After all groups have finished placing their gloves in front of the various products, review and discuss the different decisions made by each group (if any). Review each product and state the appropriate type(s) of gloves that should be used when working with that material.

Debriefing/Discussion Questions:

How did seeing a variety of types of PPE make this activity easy or more difficult?

What types of PPE or tools were missing?

What did you learn that you can take back to your workplace?

Copyright © 2006 SafetyFUNdamentals™

Mega Memory

Objectives	To make associations between class concepts and information learned
Good For	Any topic
Cautions	Reading is required
Class Length	Any
Audience	Anyone
Where it can be done	Anywhere with adequate floor space
Team Size	2-4
Time Needed	15-20 minutes
Set-Up/Materials	Pre-made question cards

Copyright © 2006 SafetyFUNdamentals™

Instructions for Mega Memory

Before class, make up 20 or more memory cards with questions so that pairs of cards have the same answer. Make the "cards" oversize by writing them on 8.5" x 11" paper. The pre-made memory cards should be placed face down randomly around the floor. Trainees should stand around the game and should be put into teams of 2 (3 or 4 for bigger classes). One team should be selected to go first. Like regular memory games, the team should pick a pair of cards to turn over. Since there are questions on the reverse instead of pictures or words, each team must find two cards that have questions with the same answers. For example, if the first card picked says "one way to prevent slips and trips" and the other card says "important in the prevention of accidental fires", and the answer (housekeeping) is the same for both, then the team has a match and can keep the cards. They also get another turn and can continue until they cannot find a match. After each turn, the cards are placed face down again. If they cannot find a match, then the next team gets to take a turn. When all of the cards are gone (or most - see note below), the game is over. The team with the most cards is declared Mega Memory Champions.

Note: Sometimes a team will be able to present as good argument that two cards have the same answer even though that was not the intended purpose. It is okay to let the team match these cards. You will not end up with a perfect number of matches in the end and a few cards may be left, but the learning aspect of the game is still at work and the competition still exists.

Debriefing/Discussion Questions:

How did this activity help you to remember the information learned in class?

What additional questions can you think of that could be a match for the cards remaining at the end of the game?

Copyright © 2006 SafetyFUNdamentals™

Message In A Bottle

Objectives	To focus on a particularly important point made during training
Good For	Any topic; as a closing exercise
Cautions	None
Class Length	Any
Audience	Any
Where it can be done	Anywhere
Team Size	Individual activity
Time Needed	5 minutes
Set-Up/Materials	Small bottles or envelopes and small slips of paper

Copyright © 2006 SafetyFUNdamentals™

Instructions for Message in a Bottle

Distribute one small bottle to each trainee. (If bottles are not available, give each trainee an envelope and tell them to imagine that this is their bottle). Tell them to pretend that they are stuck on a deserted island and since they care so much about safety in their plant, they want to send one important safety message back to their co-workers. (If the training is topic-specific, have the bottle message relate to the topic.) Collect all of the bottles and place them in a large container. Ask each trainee to select a bottle and read the message and think how they can apply it to their job. Alternatively, the bottles can be placed around the facility by the trainees for co-workers, not attending the training class, to find. In this case, a note should be included on the small slip of paper that instructs the trainee to return the bottle and message to the safety department (or you) for a prize.

Debriefing/Discussion Questions:

How did you select the message you wrote as "most important?"

If the bottle will be left in the plant for others to find, why did you select the message that you did?

Copyright © 2006 SafetyFUNdamentals™

Million Dollar Madness

Objectives	To brainstorm safety improvements
Good For	Opening activity
Cautions	None
Class Length	Any
Audience	Any
Where it can be done	Anywhere
Team Size	1-4
Time Needed	Approximately 10 minutes
Set-Up/Materials	Optional: bag of fake money or bag of chocolate gold coins

Copyright © 2006 SafetyFUNdamentals™

Instructions for Million Dollar Madness

Divide the class into teams. If there are different departments represented in the class it might be a good exercise if people from the same departments or work areas were put on the same team. Tell the class that the OSHA fairy (or other anonymous donor) dropped off a bag with 100 million dollars on the company doorstep last night with the stipulation that the money be used only for safety improvements. With money not an issue, each team has five minutes to develop a list of improvements they would make or new equipment they would buy using that money that would directly improve plant safety. Remind the class that money is no object. After all teams are finished, ask the teams to share their wish list with the class.

Note: It is a good idea to have someone write all the ideas down (or you can do it on a flip chart) so that all serious ideas can be saved and shared with management for possible consideration.

Debriefing/Discussion Questions:

How did the fact that you have unlimited funds change your thinking?

How can you make some of these ideas come true?

Which ideas surprised you as being more or less expensive than you expected?

Copyright © 2006 SafetyFUNdamentals™

Mnemonic Mania

Objectives	To develop methods to help remember important safety information
Good For	Any topic
Cautions	None
Class Length	Any
Audience	Anyone
Where it can be done	Anywhere
Team Size	Any
Time Needed	15-20 minutes
Set-Up/Materials	Nothing

Copyright © 2006 SafetyFUNdamentals™

Instructions for Mnemonic Mania

A Mnemonic is a method to help make it easier to remember something. A mnemonic can be a short rhyme, phrase, or other mental technique. For example, many people are familiar with "righty tighty, left loosey" which is a mnemonic to help remember which way to turn a screw to tighten or loosen it. Some people are taught to remember the number of days in a month by counting on their knuckles. Even "LOTO" is a mnemonic for remembering the steps in an energy control procedure (Lock-Out-Tag-Out.

Divide the class into teams and ask, after explaining what a mnemonic is and stating examples, if anyone else knows of any others. Tell their class that their job is to think of a mnemonic that relates to the class topic. This can be as simple as making a word up of the first letters of the steps of a process (like LOTO) or making up a rhyme, rap, song, poem or anything else that can help them to remember something related to the topic.

Debriefing/Discussion Questions:

What were their favorite new mnemonics heard in class and why are they favorites over others?

How will mnemonics help you to remember the topic?

How can you use this back in the workplace?

Copyright © 2006 SafetyFUNdamentals™

Mr. Safety Spud

Objectives	To use creativity in recognizing and protecting hazards
Good For	Reviewing hazards
Cautions	This can get messy; Open minded trainees required!
Class Length	At least one hour
Audience	Any
Where it can be done	Classroom
Team Size	Any
Time Needed	25 minutes
Set-Up/Materials	Potato for each trainee or team and miscellaneous "building" supplies such as tissues, binder clips, pushpins, tape, etc. and "hazard papers" placed into a box

Copyright © 2006 SafetyFUNdamentals™

Instructions for Mr. Safety Spud

Before class, write out various workplace hazards on slips of paper and place them into a box. Divide the class into teams and give each team a potato as well as an assortment of office supplies such as pushpins, tacks, paperclips, tissues, post-it notes, binder clips, etc. If possible, bring in additional supplies from the cafeteria or home such as toothpicks, aluminum foil, plastic wrap, plastic utensils, a trash bag, scissors, tape, etc. Let each team pick from the prepared "box of hazards." Tell each team that they each have 15 minutes to complete their potato heads. They should use any of the supplies found in the room to highlight the parts of Mr. Potato Head that are at risk for their team's selected hazard and if they have time, they can work for bonus points by designing PPE for those parts - all out of materials found in the room. For example, if one of the hazards was glass handling, the team could make a hand out of any of the office supplies and for bonus points, they could use tissues or post-it notes to make a glove. Tell the teams that they will receive one point for each correctly identified hazard *that they were able to make on Mr. Safety Spud* and two bonus points for each type of PPE they identify and *place on Mr. Safety Spud*.

Debriefing/Discussion Questions:

What were the benefits of having materials to make specific PPE?

Did you find this exercise easy or hard and why?

What did you learn that you can take back to the workplace?

Copyright © 2006 SafetyFUNdamentals™

Mystery Injury

Objectives	To develop good question asking practices
Good For	Accident investigation training
Cautions	Do not imitate any actual injuries or accidents that occurred in the facility
Class Length	At least 45 minutes
Audience	Any
Where it can be done	Anywhere
Team Size	At least ten trainees
Time Needed	Fifteen minutes
Set-Up/Materials	Prepared injury/accident cards

Copyright © 2006 SafetyFUNdamentals™

Instructions for Mystery Injury

Before class prepare injury/accident cards by writing the description of an injury on an index card. Divide the class into pairs. Give one member of each pair an injury/accident card. The trainees with cards are the "injured" employees, or the ones showing up in the medical office with a complaint. Each card will show an injury, as well as a short description of how it happened. The "interviewer" must ask questions to the "injured" employee to discover how the injury occurred. The "injured" is not to come right out and tell the interviewer how the injury occurred but rather they should be as vague as possible so that the interviewer can practice their accident interviewing techniques as much as possible.

After about 10 minutes, and if time allows, give a new injury/accident card to the "Interviewer" and have them switch places with the "Injured." Repeat the activity.

Debriefing/Discussion Questions:

What did the injured specifically do that made it especially easy or difficult to discover how the injury occurred?

What could you have done better as an interviewer?

What did you learn that you will take back to the job?

What will you do differently the next time you complete an accident investigation?

Copyright © 2006 SafetyFUNdamentals™

Mystery Muck

Objectives	To practice good questioning techniques
Good For	Hazardous material training or MSDS review
Cautions	Reading required for version 1
Class Length	Any
Audience	Any
Where it can be done	Anywhere
Team Size	2-4
Time Needed	About 10 minutes
Set-Up/Materials	Various harmless "mucky" substances and the MSDS for each

Copyright © 2006 SafetyFUNdamentals™

Instructions for Mystery Muck

Collect various boxes, cans, bottles or other containers and place an unknown but harmless substance in each. Make sure you have an MSDS for each. Divide the class into teams and tell the class that they have been sent to clean up the mystery muck that has been found but they must first identify it. They can ask five Yes and No questions before they must make a guess. Good items to include as "muck" are Silly Putty®, "Slime" from any toy store, a mixture of oatmeal or other cereal and water, mud, or anything else that you can come up with. If it is safe, you can actually use a material found in the workplace.

For version 1, you can give each team of two a container of "muck" plus the associated MSDS and have each person in the team take on one role (either investigator or MSDS holder). In version 2, you can give each team a container of muck and you as facilitator can hold onto all MSDS and answer all questions.

Debriefing/Discussion Questions:

What would have made this exercise easier?

Did it help or hurt to have the actual substance in front of you and why?

What additional questions do you wish you asked?

Copyright © 2006 SafetyFUNdamentals™

No Danger Darts

Objectives	To review class material
Good For	Any topic
Cautions	Do not use real darts!
Class Length	At least 45 minutes
Audience	Any
Where it can be done	Classroom
Team Size	2-5
Time Needed	About 15 minutes
Set-Up/Materials	Magnetic or foam dart board; prewritten questions of increasing difficulty pertaining to each color on the board

Copyright © 2006 SafetyFUNdamentals™

Instructions for No Danger Darts

Before class begins, write out questions relating to the class material on colored cards that match the colors on the dart board (do not use real darts for this activity). The questions should be easy, medium and hard in accordance with the points available for each color on the dart board. Using one of the team-sorting methods described in this book, place the class into teams of 2-5 and decide which team will go first, second, and so on.

Have a representative from each team stand a few feet from the dart board. When you say go, have each representative throw their "dart." After all darts have landed, ask each team a question of the corresponding color and award points as earned. After 5 or so rounds (depending on time available), tally all points and announce the winner. A different team member should take a turn each round.

Debriefing/Discussion Questions:

What was the most important thing you remembered after completing this exercise?

What was your strategy for collecting the most points for your team?

What was the most difficult question and why?

Copyright © 2006 SafetyFUNdamentals™

No Dice

Objectives	To review class materials
Good For	Any topic
Cautions	Some companies or individuals may oppose the use of dice at all so check first
Class Length	At least 30 minutes
Audience	Any
Where it can be done	Anywhere but outdoors may be tricky
Team Size	2-4
Time Needed	Approximately 20 minutes
Set-Up/Materials	Large foam dice are preferred (but smaller typical dice will also work)

Copyright © 2006 SafetyFUNdamentals™

Instructions for No Dice

Before the class begins, write down 3 columns of review questions numbered 1-6 on flip chart paper or on a slide if it can be projected larger. If there is a lot of class material and more questions are needed, make additional columns of six questions each but remember they must all be different questions. When it is time to review, give each team of 2-4 people a pair of dice. Teams of 4 work nicely. Each team takes turns rolling the dice. The value on the face of one die represents which question they must answer and the value of the other die represents how many points they will receive if they answer that question correctly. Each team can select which die will represent each but generally, the die with the highest value will be used for the points. Each question can be answered only once so if a team rolls a 6 and a 1, and all questions numbered 1 have been answered already, they will have to answer question #6 for 1 point. The teams take turns rolling the dice and answering questions until all questions have been answered. If a team answers incorrectly, they lose their turn but the question still remains available. If a team rolls two numbers and both of those numbers (in all columns) have been answered, they also lose a turn and the dice are passed to the next team. At the end of the game, all points are tallied and the winning team announced.

Debriefing/Discussion Questions:

What did you learn or remember about the topic?

How did competition affect your ability to answer the questions?

Copyright © 2006 SafetyFUNdamentals™

No More Excuses

Objectives	To share ideas to improve safety-related communication
Good For	Topics where compliance is usually a problem (such as with the wearing of PPE)
Cautions	Caution the trainees not to use the real names of co-workers
Class Length	At least 30 minutes
Audience	Supervisors
Where it can be done	Anywhere
Team Size	Individual/class exercise
Time Needed	At least 10 minutes
Set-Up/Materials	Paper, small box

Copyright © 2006 SafetyFUNdamentals™

Instructions for No More Excuses

Ask each trainee to write down one excuse they often hear related to compliance with safety rules or procedures. The excuse should relate to the topic being presented. Collect all of the papers and put them into a box. One by one, pull an excuse out of the box and read it to the class. Ask the class to shout out the feedback or response they would give to someone who presented that excuse for not being safe. After they have responded to that excuse, rip up the paper and toss little pieces high in the air. (For less mess but also less visual impact, crumple the excuse and throw it into the trash can). For longer classes, you might want to pull two or three excuses at a time and then occasionally return to the excuses after additional material has been covered. When the last excuse has been read and destroyed, announce "No More Excuses."

Debriefing/Discussion Questions:

Which excuses were ones you hadn't heard before?

Which excuses do you think were/are legitimate and why?

What was the best response you heard given to an excuse?

Copyright © 2006 SafetyFUNdamentals™

Pass The Hard Hat

Objectives	To encourage anonymous question asking
Good For	Any topic
Cautions	Trainees need to be able to read and write
Class Length	Any
Audience	Any
Where it can be done	Anywhere
Team Size	Larger is better
Time Needed	2 - 3 minutes per trainee
Set-Up/Materials	Small slips of paper, pens, hardhat, timer

Copyright © 2006 SafetyFUNdamentals™

Instructions for Pass the Hard Hat

For shorter review classes, ask each attendee at the beginning of the class to write one question about the topic on a slip of paper and fold it up. After everyone is finished, pass the hardhat around and have everyone place their paper into the hat. When the hat has gone around once, have the hat passed around again but this time each person is to take out one slip of paper (not their own) and read it. Tell the team that they each have 3 minutes to find the correct answer to their new question. They can work on it themselves or ask other trainees in the class but they only have three minutes. At the end of three minutes, go around the room and ask trainees to read their question and the answer. If the class is large, you can ask for volunteers or you can select trainees randomly such as every 3rd person. This exercise can make up the entire class for topics that the class is very familiar with. If it is used with new material, it can be done after a break or at the end of class.

Debriefing/Discussion Questions:

Was it easy or hard to think of a good question and why?

What did you remember that you had forgotten?

What information can you use back on the job?

Copyright © 2006 SafetyFUNdamentals™

Pick a Card - Any Card

Objectives	Initial determination of class knowledge
Good For	Topics that cover a lot of information
Cautions	Some individuals or companies do not look favorably upon card games
Class Length	At least 45 minutes
Audience	Any but reading is required
Where it can be done	Classroom setting or somewhere with tables
Team Size	Any
Time Needed	Approximately 15 minutes
Set-Up/Materials	A list of 52 prewritten questions (many questions can be similar), a deck of cards for each pair of students.

Copyright © 2006 SafetyFUNdamentals™

Instructions for Pick a Card - Any Card

Divide the class into pairs and give each team a deck of cards (Jokers removed) along with a list of prepared questions about the topic. There should be 4 columns of 13 questions for a total of 52 questions. Each column represents a different suit in the cards (Hearts, Diamonds, Spades, and Clubs). Each trainee should start out with 5 cards that are hidden from their partner's view. Tell the class that the object is for each player to get his or her cards into pairs. When they have two of the same card, either suit or value, they can put the pair down on the table. Each trainee should try to get all of their cards on the table before the partner. Similar to "Go Fish", one player asks the other if they have any of a particular card in their hand. If they do, they need to give it to the other player. The other player will now have a pair and can put the pair down. If they do not have the card that was requested, the person who asked must answer the corresponding numbered question from the question list that was distributed. If they get the question right, they pick one card from the pile. If they are unable to answer the question or get it wrong, they must pick two new cards from the pile and the other player now gets a turn. Play continues until one of the players has gotten rid of all of his or her cards.

Note: face cards are valued as follows: Jack - #11, Queen - #12 , King - #13 and Ace - #1. All other cards have the values shown.

Debriefing/Discussion Questions:

What was the strategy you used to get all of your cards on the table first?

What was the most difficult question you had to answer and why?

Copyright © 2006 SafetyFUNdamentals™

Picture Perfect

Objectives	To review key points of the training class
Good For	Any topic
Cautions	Be sure everyone clearly understands what a pictogram should look like
Class Length	Any
Audience	Any
Where it can be done	Anywhere
Team Size	Individual
Time Needed	Approximately 10 minutes
Set-Up/Materials	Paper and markers or crayons

Copyright © 2006 SafetyFUNdamentals™

Instructions for Picture Perfect

Explain to the class what a pictogram is and then ask the class members to draw a pictogram representing a topic covered in the training class. Tell the trainees that they should draw the pictogram as if the people that had to read it did not know any English (or how to read any language). After each trainee has completed a pictogram, collect all of the papers. Mix them up and then redistribute them to the trainees so that they do not have their own pictogram. Go around the room and ask each person to show the pictogram they received and explain what it means to the class.

Debriefing/Discussion Questions:

How did not being able to use words make you think differently?

Why was it difficult to explain someone else's pictogram?

How could you use your pictogram back in the workplace?

Copyright © 2006 SafetyFUNdamentals™

Pizza Pie

Objectives	To demonstrate how different departments work together
Good For	Any topic
Cautions	None
Class Length	Any
Audience	Any
Where it can be done	Classroom
Team Size	Variable but 8 teams are needed
Time Needed	Approximately 10-15 minutes
Set-Up/Materials	Large poster board or cardboard first cut into a circle and then into pizza shaped wedges

Copyright © 2006 SafetyFUNdamentals™

Instructions for Pizza Pie

Before the class begins, get a large piece of paper (poster board would be better) and draw a large "pizza." Divide and cut the pizza into 8 sections (or slices). Write different company departments around the edges (or crust) of each piece, such as Facilities, HR, Medical, Safety, Environmental, Union, Accounting, Purchasing, Operations, Quality, etc. Divide the class into teams. Try to have at least several people on each team. Give a "slice" to each team. If you have a smaller class, give each team two slices. Tell the team that they need to list at least 5 things on their slice that the listed department must do to ensure or encourage safety related to the topic you are presenting. Tell the trainees that they should tape their slice of pie to the board or wall in the front of the room when they are finished. (Pieces should be taped back into the "pie" shape).

Debriefing/Discussion Questions:

Was it easy or difficult to think about other department's responsibilities for safety and why?

What did you learn that will help you when interacting with other departments?

What misunderstandings did you become aware of?

Copyright © 2006 SafetyFUNdamentals™

safetyFUNdamentals

Plain English

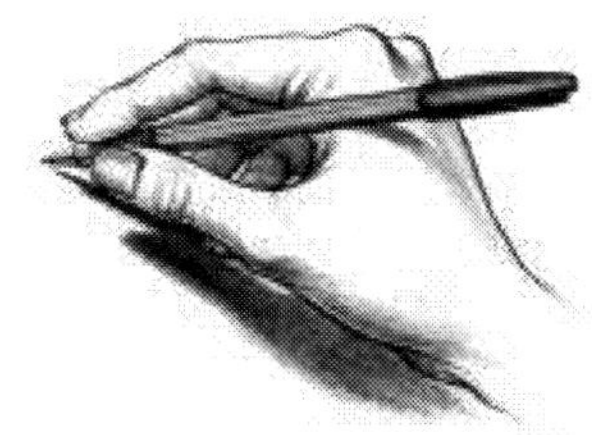

Objectives	To understand how to read and understand OSHA regulations
Good For	OSHA-required training
Cautions	Good reading and writing skills required
Class Length	At least 30 minutes
Audience	Any
Where it can be done	Anywhere but classroom preferred
Team Size	3-5
Time Needed	Depending on the standard, 10-20 minutes
Set-Up/Materials	Copies of related regulations and paper and pencils

Copyright © 2006 SafetyFUNdamentals™

Instructions for Plain English

Provide each team with a copy of a section of an OSHA regulation (or other regulation such as those issued by EPA, MSHA, States, etc.) that pertains to the topic you are presenting. Ask each team to decipher what they read and write a summary of that particular part of the regulation as a plain English paragraph. Tell them they cannot use any words with more than 6 letters and all sentences must have less than 10 words. At the end of the exercise, they will share their plain English paragraph with the rest of the class.

Debriefing/Discussion Questions:

How did you translate the regulation you were given into something much easier to read?

What was the most difficult part of this exercise?

If you worked for OSHA, or other regulatory agency, what would you add to this regulation and why?

Copyright © 2006 SafetyFUNdamentals™

safetyFUNdamentals

Plot the PPE

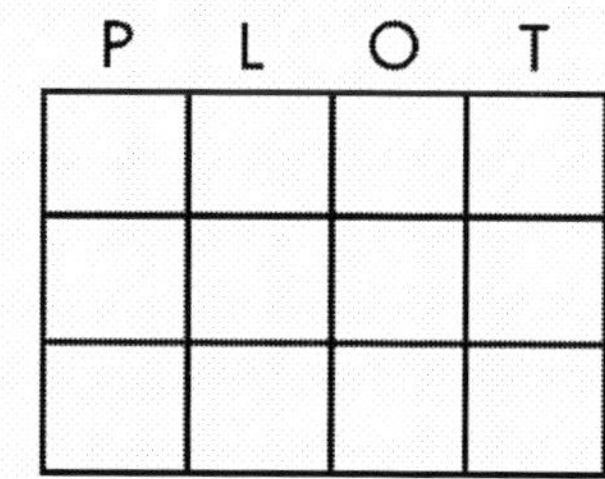

Objectives	Reviewing Information
Good For	Topics with a lot of content
Cautions	Reading and writing required
Class Length	Any
Audience	Any
Where it can be done	Anywhere
Team Size	4 - 5
Time Needed	Approximately 10 minutes
Set-Up/Materials	A "plot" paper for each team, pencils

Copyright © 2006 SafetyFUNdamentals™

Instructions for Plot the PPE

Make a 3x4 matrix for each team. On the short side, write three categories of personal protective equipment such as goggles, a respirator and gloves. On the other side, write the letters P.L.O.T. Divide the class into teams of 3-5. Tell each team they have seven minutes to think up as many items as they can that relate to the type of PPE on each line that begins with the corresponding letter (P, L, O, or T). At the end of the time, give each team one point for each item listed. Award a bonus point for any item listed that does not appear on any other team's plot. Announce and reward the winning team and then debrief the activity. A sample PLOT matrix is available in Appendix H.

Note: This exercise can be use to "plot" anything. PPE types were only used here as one example.

Debriefing/Discussion Questions:

How did this exercise make you think differently about the three types of PPE?

How did working in a team help you?

After hearing other teams' answers, what new information did you learn?

Copyright © 2006 SafetyFUNdamentals™

Pre-test Protest

Objectives	Review of previously learned information
Good For	Annual refresher courses
Cautions	Reading and writing required
Class Length	Any
Audience	Any
Where it can be done	Anywhere
Team Size	3-4
Time Needed	Approximately 10 minutes
Set-Up/Materials	Paper for each team, pencils and pens

Copyright © 2006 SafetyFUNdamentals™

Instructions for Pre-Test Protest

At the beginning of class, put the trainees into teams of 3 or 4. Tell the trainees that each team needs to think of 5 key things that are important to learn from the class. They should write these things in the form of a question and they should write their five questions on a piece of paper. After 5 minutes, have the teams switch papers. Announce that they have just written the pre-test for the class and each team should work together to answer the 5 questions they were given. After approximately 5 minutes, collect the papers and review them as a group.

Note: This activity is most effective for annual refresher courses where the trainees already have an understanding of the material and experience applying it in the workplace.

Debriefing/Discussion Questions:

How did your team work to come up with 5 questions?

How did this exercise help you to realize you may have forgotten some key areas from previous training?

Which questions did you find most difficult?

Can you think of any other questions that should have been included that weren't and if so, what were they?

Copyright © 2006 SafetyFUNdamentals™

Quiz Wiz

Objectives	To encourage trainees to focus on the most important concepts presented in a training class
Good For	Any topic
Cautions	Writing required
Class Length	At least 30 minutes
Audience	Any
Where it can be done	Anywhere
Team Size	2-4
Time Needed	Approximately 10-15 minutes; can be spread out throughout the class
Set-Up/Materials	Paper and pens/pencils

Copyright © 2006 SafetyFUNdamentals™

Instructions for Quiz Wiz

In this activity, trainees will be making up their own class quiz. If a training class has several modules or sections, the Quiz Wiz Activity can be done in parts, after each section is completed. If there is only one main section, you can also do this activity after all of the class information has been presented.

If you have, for example, five different subjects in your safety training class, you would pause after each section and ask the trainees to work in pairs to develop two quiz questions related to that topic. At the end of the class, collect all of the questions and redistribute them so that each pair has a new (peer-developed) quiz to work on together. When all teams have finished, ask if anyone has questions about any of the quiz questions and review them as a class.

You can also do this activity as a class by asking the class as a whole to come up with two great quiz questions related to the content. When the class has agreed, write these questions on a flip chart page that you will keep coming back to after each section. When the class is over, display the full list of questions and ask the class to answer them.

Debriefing/Discussion Questions:

Was it easy or difficult to make up your own quiz questions and why?

Did you write your questions differently knowing that you may have to answer them yourself?

What was the most difficult question and why?

Copyright © 2006 SafetyFUNdamentals™

Real World Respiration

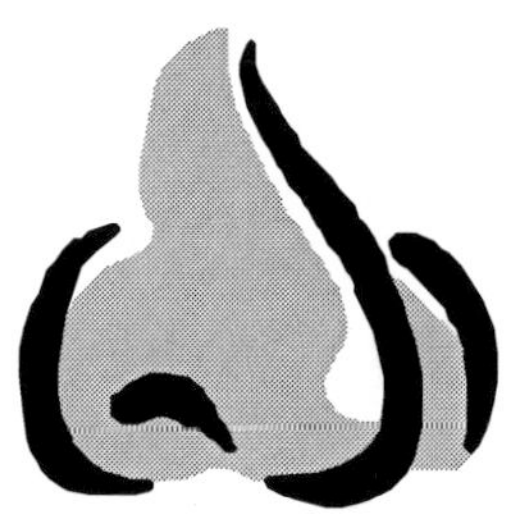

Objectives	To demonstrate hands-on understanding of respiratory protection
Good For	Respiratory Protection topics
Cautions	None
Class Length	At least 30 minutes
Audience	Any
Where it can be done	Classroom
Team Size	3-5
Time Needed	Approximately 15 minutes
Set-Up/Materials	Miscellaneous "building" items (see next page)

Copyright © 2006 SafetyFUNdamentals™

Instructions for Real World Respiration

Collect various miscellaneous "building items" such as balloons, marbles, straws, paper, pipe cleaners, envelopes, plastic spoons, clay - anything safe you have access to that could possibly be used to develop a model. After putting the class into teams of 3-5, ask the teams to simulate a portion of the training such as how the lungs are structured or how they function in the human body, including how particles of different sizes deposit in different parts of the respiratory system. If space and time allow, let the trainees leave the room to collect other things that would be helpful in building their model. After a sufficient amount of time, have each team present and explain their model.

Debriefing/Discussion Questions:

Why did you find this exercise difficult or easy?

How did you use your training materials or class notes in preparing your model?

What part of this activity will help you the most in remembering the key points of respiratory protection?

Copyright © 2006 SafetyFUNdamentals™

Rock, Paper, Scissors

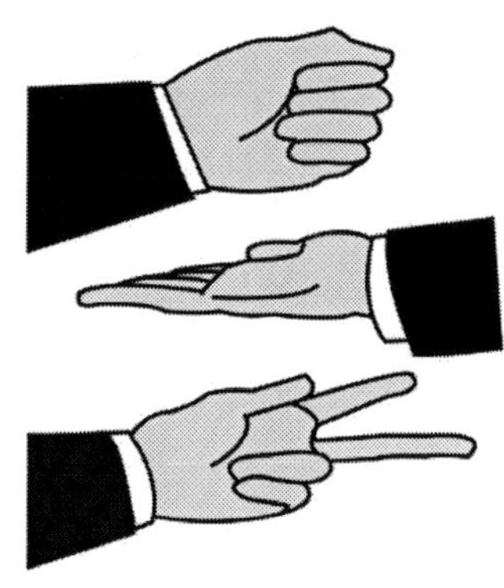

Objectives	To encourage trainees to think about different solutions to problems
Good For	Any topic
Cautions	Explain game thoroughly
Class Length	Any
Audience	Any
Where it can be done	Anywhere
Team Size	Teams of two
Time Needed	From 2 - 10 minutes
Set-Up/Materials	Prepared "condition cards" (a list of safety hazards or other related issues)

Copyright © 2006 SafetyFUNdamentals™

Instructions for Rock, Paper, Scissors

Ask if anyone in the class is familiar with the game "Rock, Paper, Scissors." Have two volunteers pick a partner and demonstrate the game to the class. Explain to the class that in this game, "rock" stands for discipline, "paper" stands for written procedures and policies and "scissors" stand for engineering controls. Have all trainees stand and form teams of two. Give each team at least 6 condition cards. It is okay if all pairs get the same set of 6 cards.

Have each team turn over a condition card. They have a few seconds to think about the best way to address the condition or situation either through rock (discipline), paper (procedures or policy), or scissors (engineering controls). Just like in a normal rock, paper, scissors game, they need to first say "Rock, Paper, Scissors - Shoot" and when they do, they need to display their choice. As in the real game of "Rock, Paper, Scissors," "rock" beats "scissors," "scissors" beats "paper" and "paper" beats "rock." The teams should keep score of how many wins each person has along with the number of ties but the real value in this exercise is from each member of the team seeing how the other person would handle the situation if only three options were available (discipline, policies, or engineering controls). If they both "shoot" the same thing, it is a tie and means that they agree on how the situation should be handled. After all of the condition cards have been used, the team can sit down.

When all teams are finished and seated, ask for the number of "ties" each group experienced and discuss.

Debriefing/Discussion Questions:

Did you think you would "tie" more or less with your partner? Why?

Did you ever change your mind after hearing the other person's argument for what they chose (rock, paper, or scissors)?

Copyright © 2006 SafetyFUNdamentals™

Rotten Rhymes

Objectives	To teach trainees methods of remembering key safety information
Good For	Topics with steps which must be completed in entirety or in a specific order
Cautions	Inappropriate responses may be presented
Class Length	At least 30 minutes
Audience	Any
Where it can be done	Anywhere
Team Size	2-4
Time Needed	Approximately 10 minutes
Set-Up/Materials	Paper and Pencils or Pens

Copyright © 2006 SafetyFUNdamentals™

Instructions for Rotten Rhymes

Ask the trainees to create a rhyme based on the information they are learning (this also works well for any topic area but remember to have the class work in teams). Review what a mnemonic is if necessary (see Mnemonic Mania Activity).

If anyone gets stuck, provide the following beginnings...

There once was a man from ____________________
Who always went ____________________________
Until one day he _____________________________
And his children_______________________________
And his friends ______________________________

Debriefing/Discussion Questions:

How can a rhyme or mnemonic that you create help you to remember important safety information?

What was your favorite rhyme or mnemonic that you heard and why?

How will you use your favorite rhyme or mnemonic back in the workplace?

Copyright © 2006 SafetyFUNdamentals™

Safety-Libs

Objectives	To review content in a humorous way
Good For	Any topic; opening activity
Cautions	Make sure trainees understand what is meant by "verb or action word"
Class Length	At least 30 minutes
Audience	Any
Where it can be done	Classroom
Team Size	Any
Time Needed	5-10 minutes
Set-Up/Materials	Prepared safety-libs, pens/pencils

Copyright © 2006 SafetyFUNdamentals™

Instructions for Safety-Libs

Explain that you have a safety case study in front of you and you need help filling in the blanks. You will first ask for different types of words which anyone in the class can shout out. (Write these words in the appropriate spaces). When you are finished, read the final story to the class (it should be funny). Now, go back through the story again slowly, but this time read the entire sentence and pause before each class-offered word and ask the class to offer a possible correct substitution.

Writing your own Safety-Lib is easy. Just start with a short story and then go back and remove several action words and safety-related nouns. You can also use a safety-related story from a newspaper or other report and take out similar words. An example Safety-lib form can be found in Appendix J.

Debriefing/Discussion Questions:

Why do you think hearing something funny can help you to learn?

What information can you use back in the workplace?

Copyright © 2006 SafetyFUNdamentals™

Safety Consultant For a Minute

Objectives	To solicit feedback and advice from peers
Good For	Anything but broader topics like safety leadership work especially well
Cautions	Poor readers and writers may have difficulty
Class Length	Any
Audience	Any
Where it can be done	Anywhere but works best with employees seated with a table available for writing
Team Size	Any but a minimum of 6 is preferred
Time Needed	Approximately 8 minutes per trainee
Set-Up/Materials	Paper and Pencils and timer

Copyright © 2006 SafetyFUNdamentals™

Instructions for Safety Consultant for a Minute

If the class is large, divide the trainees into smaller teams of 6 to 8. Give each trainee a piece of blank paper and ask them to write a question across the top about a situation where they would want to receive advice or suggestions. Tell the trainees that when you say go (or ring a bell or blow a whistle or make some other noise), they should pass their paper to their right (within their smaller team), and then begin to answer the question on the top of the paper they just received. Emphasize that they need to read it and answer it quickly because they only have one minute before you tell them to stop and pass the papers to the right again. When they receive the second paper, they need to quickly read over the answer that was already provided so that they do not duplicate what was already written. Once again, start and stop the class after one minute. Continue to have the trainees pass the papers to their right until all papers have returned to their original owner. Give the trainees some time to review all of the suggestions and solutions that were written in response to their question.

Debriefing/Discussion Questions:

Why did you find this activity easy or hard?

Was it more difficult to work under pressure? Why?

What makes a quality answer or solution?

Is it easier to solve someone else's problem or you own?

Copyright © 2006 SafetyFUNdamentals™

Safety Consultant For a Minute

Alternate Version

Objectives	To solicit feedback and advice from peers
Good For	Any topic
Cautions	Poor readers and writers may have ditticulty
Class Length	Any
Audience	Any
Where it can be done	Anywhere but works best with employees seated with a table available for writing
Team Size	Any but a minimum of 6 is preferred
Time Needed	Approximately 8 minutes per trainee
Set-Up/Materials	Paper and Pencils, timer and fake money or poker chips

Copyright © 2006 SafetyFUNdamentals™

Instructions for Safety Consultant for a Minute

Alternate Version

Follow the steps for "Safety Consultant for a Minute" from the previous activity but also include the following steps.

Give each trainee ten fake $1 bills or ten poker chips each. Tell them that they need to give out all of their money or chips to all of the participants on their team based on the value of the answer they received. For example, they could give all of their $10 to one person if the received an answer that was so wonderful it would solve all of their problems or they could give one person $5 and split the remaining $1bills between the others. (For this to work, each trainee should be told to write their initials next to their answers as the sheets go around so that everyone knows whose answers belong to whom). In the end, see who in the team has the most money and who out of the entire class has the most money. Ask the winners (there will likely be ties) to share the question they answered that provided the windfall and the solution or suggestion they provided.

Debriefing/Discussion Questions:

Why did you find this activity easy or hard?

Was it more difficult to work under pressure? Why?

How did you dete rmine who to give your money to?

What makes a quality answer or solution?

Copyright © 2006 SafetyFUNdamentals™

Safety Hangman

Objectives	To review keywords associated with a topic
Good For	A quick activity when returning from breaks
Cautions	Good spelling ability is helpful
Class Length	Better for longer classes that include short breaks
Audience	Any
Where it can be done	Anywhere
Team Size	3-4
Time Needed	5 minutes
Set-Up/Materials	Hangman sheets, prepared word cards, pens/pencils

Copyright © 2006 SafetyFUNdamentals™

Instructions for Safety Hangman

Before the class, write key safety words and phrases from the class onto folded index cards with the "secret" words written on the inside. Divide the class into teams of 3-4. One person should be given the folded card with the secret safety word and the other team members will take turns guessing letters and trying to solve the puzzle. Explain how "hangman" is played. Explain that Safety Hangman is very similar but instead of body parts (the body is already hanging in this game), each correct guess results in a piece of personal protective equipment being removed from the "hangman". (A piece of PPE can be removed by simply crossing it out.) When all pieces of PPE are removed from the hangman, the game is over. A sample safety hangman can be found in Appendix I. If you use the sample provided each team will have 8 chances to guess the correct answer. If you draw your own, as many possible pieces of PPE should be included on the hangman to maximize the chances the team has to guess the secret word.

Debriefing/Discussion Questions:

How did knowing you only had a limited number of chances effect your guessing?

How could this concept (a limited number of chances) be applied to safety?

By knowing that the word to be guessed was covered in the class material, how could this help you to remember the content?

Copyright © 2006 SafetyFUNdamentals™

Safety Math

Objectives	To help trainees to make connections with class information
Good For	Warm-up/Preview activity or when returning from a break
Cautions	Minimal reading and writing required
Class Length	Any
Audience	Any
Where it can be done	Anywhere
Team Size	2-3
Time Needed	5 - 10 minutes
Set-Up/Materials	Prepared Safety Math problems

Copyright © 2006 SafetyFUNdamentals™

Instructions for Safety Math

Write up a page of 15 - 20 "safety math" problems and distribute to the class at the beginning as either a preview or leave on desks for trainees to work on when they come back from a break. To create Safety Math problems, think of two situations or objects that could lead to a third condition or action and write these up as a math problem would look. For example:

Water + Electricity = ______________ (We all know the answer is shock)

Similarly, try: No Handrail + Rain = ______________.

You can also provide the "sum" but not all the components such as: Back bent + ______________ = back sprain

Instruct the trainees to work on the problems in groups of 2 or 3. After every group has completed the problems, review them as a class and discuss.

Debriefing/Discussion Questions:

Were you surprised at how much information you remembered? Why?

What problems did you have trouble with? Why?

Try to make up your own safety math problem and share it with the class.

Copyright © 2006 SafetyFUNdamentals™

Safety Menu

Objectives	To practice prioritizing safety hazards or corrective actions
Good For	Hazard recognition and corrective action classes
Cautions	Reading skills required
Class Length	Any
Audience	Any
Where it can be done	Anywhere
Team Size	2-4
Time Needed	Approximately 15-20 minutes
Set-Up/Materials	Prepared Safety Menus

Copyright © 2006 SafetyFUNdamentals™

Instructions for Safety Menu

Divide the class up into teams of 2 - 4. Give each team of prepared Safety Menu. (The menus show different pieces of safety related equipment and prices). Give each team an imaginary $5000 dollars to spend (use paper pretend money if available). Present each team with a scenario that requires the use of safety equipment or procedures. Tell them that they must spend their money wisely in order to protect as many workers as possible who may be performing that task. They cannot go over the amount of money they were given.

After 10 - 15 minutes, stop the teams from working with their menus and ask them 1) how many workers they were able to protect and 2) what they purchased from the menu and 3) how much money they spent. As each team provides their answers, ask them about the decisions they made and encourage other teams to do the same.

Note: The $5000 is an arbitrary amount Depending on the type of menu items and the scenarios given to each team, the amount to spend could be higher or lower.

An example of a Safety Menu can be found in Appendix K .

Debriefing/Discussion Questions:

How did you make your decisions?

Was there any disagreement within the group and why?

If you were not able to purchase everything that you think that would benefit the workers in your scenario, how much more money would you need and why?

Copyright © 2006 SafetyFUNdamentals™

Safety Sequence

1➔2➔3➔4➔5

Objectives	To place a sequence of events in the proper order or to see how activities relate to one another
Good For	Any topic with connected parts; Larger classes
Cautions	Reading required
Class Length	At least 45 minutes
Audience	Any
Where it can be done	Anywhere
Team Size	Preferably at least 8 people (and 8 steps) per team
Time Needed	Five minutes
Set-Up/Materials	Pre-printed cards with one step in the process written on each card.

Copyright © 2006 SafetyFUNdamentals™

Instructions for Safety Sequence

Before the class, write out steps in a process (preferably at least 8 steps) on index cards. A duplicate set of cards will be needed for each team.

Place a pre-printed step card face down in front of each trainee in each group but make sure the cards are not in order. Tell the class that there will be a race to see which team can get themselves in the proper sequence first. When you say start, they should turn over their card and work with their team to get themselves in the right order. They should stand in a line when they are finished with the person holding step one in the first position, the second person holding step two in the second position, etc.

Debriefing/Discussion Questions:

How difficult was it for you to find the correct location for your card and why?

How did this activity show you how different parts work together with respect to this topic?

What steps or activities were left out of this exercise?

Copyright © 2006 SafetyFUNdamentals™

Safety Sort

Objectives	To encourage discussion about different types of hazards and perception of those hazards
Good For	Hazard recognition classes
Cautions	Reading required by at least one team member
Class Length	At least 60 minutes
Audience	Any
Where it can be done	Anywhere
Team Size	3-5
Time Needed	Approximately 20-30 minutes
Set-Up/Materials	Prepared hazard cards, envelopes

Copyright © 2006 SafetyFUNdamentals™

Instructions for Safety Sort

Before the class, write out 30 different class-related safety hazards, including behaviors, onto 30 index cards. Since you will need several sets of these cards, it is easiest to type them on a computer and print them out as labels with a set for each team.

Divide the class into teams of 3-5. Give each team a prepared envelope containing 30 hazard cards. (Each team should receive the same cards in their envelopes). Ask each team to sort the cards into three piles. The first pile should include hazards, conditions or behaviors that can be addressed by any employee on the spot. The second pile should be hazards that require a work order or major purchase to be made and the third pile should be hazards that require immediate work stoppage.

After all teams have had time to sort their cards into three piles, ask for one team to volunteer to share the items in their piles with the class. After they have finished reading through their piles, ask the class if any other team had different groupings and discuss. (This will most likely lead to vigorous discussion). After all hazards have been discussed, continue to the debriefing.

Debriefing/Discussion Questions:

Why do you think teams had different items in different piles?

What could happen if different departments or divisions also categorized hazards differently?

What can you use from this exercise back in the workplace?

Copyright © 2006 SafetyFUNdamentals™

Safety Story-Board

Objectives	To review a safety, health or environmental topic that consists of several steps
Good For	Processes or procedures that involve several or many steps
Cautions	Writing required
Class Length	At least 45 minutes
Audience	Any
Where it can be done	Anywhere but easiest in a classroom
Team Size	Any
Time Needed	20-30 minutes depending on topic
Set-Up/Materials	8.5" x 11" pieces of cardboard, markers, masking tape

Copyright © 2006 SafetyFUNdamentals™

Instructions for Safety Storyboard

As a review, first divide teams into groups of 3 to 6. Ask each team to list the steps in the process or procedure. Place a supply of the cardboard pieces and markers on a table in the room. After each team has their list of steps, ask each to pick up the number of cardboard pieces they will need plus markers, and then draw each step on each one of the pieces of cardboard. There cannot be more than one step on a board. After time is up, ask each team to tape their storyboard to the wall in the correct order. Alternatively, you can have each team switch storyboards with another team (the story boards should not be in order). Ask the teams to put the other team's boards in the correct order.

Note: If a large amount of content is presented, you can assign different topics to different teams and then repeat the above.

Debriefing/Discussion Questions:

How did "seeing" the process help you to remember it? Why?

What did you learn new?

How can you use this process back at work?

Copyright © 2006 SafetyFUNdamentals™

Safety Strips

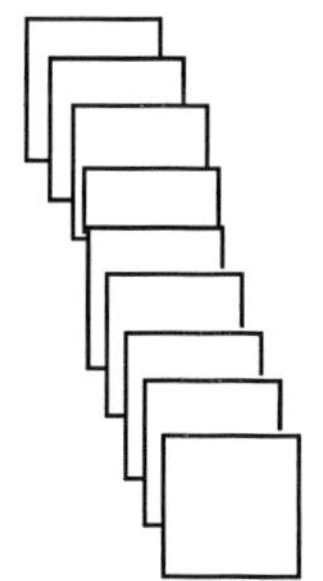

Objectives	To encourage teams to work together to recall as much of the class content as possible
Good For	Any topic
Cautions	Writing required. Also, make sure there is adequate space for the activity
Class Length	At least 30 minutes
Audience	Any
Where it can be done	Classroom
Team Size	3-4
Time Needed	Approximately 10-15 minutes
Set-Up/Materials	Pad of Post-It® notes for each trainee, pens and pencils

Copyright © 2006 SafetyFUNdamentals™

Instructions for Safety Strips

This activity serves as either a review to be used at the beginning of an annual refresher course or at the end of a class. Put the class into teams of 3-5 and give each team member a pad of post-it notes. Tell the teams that when you say go, each team must write down one key point related to the topic on a post-it note. When the first point is written, the person who wrote it must go the front of the room and place it on the wall or other flat surface. Any other key points identified should be written on post it notes and stuck (attached) at the bottom of the previous note by the person who wrote it. The goal is to see which team can build the longest strip of post-its before time runs out. There are 3 important rules to convey to the class: 1) No running, 2) only one person from each team can be out of their seat at a time, and 3) no duplication of key points by team members.

When time is up, only the post-it notes already attached at that time will be counted and the winning team should be decided based on the number of key points that are attached. Check over each strip though to make sure that there are no duplicates. If there are, simply remove that post-it and reattach the rest of the strip.

Debriefing/Discussion Questions:

Why do you think teams had different key points?

Was it easy or difficult to remember the key points and why?

What other key points did you forget to include that you wish you had remembered?

Copyright © 2006 SafetyFUNdamentals™

Scavenger Hunt

Objectives	Apply information learned in class to the workplace
Good For	PPE, Hazard Communication or any class that provides information immediately useful
Cautions	Make sure trainees understand all safety rules associated with this game
Class Length	At least 90 minutes
Audience	Any
Where it can be done	Anywhere
Team Size	3 - 5
Time Needed	Approximately 30 minutes
Set-Up/Materials	Prepared scavenger hunt list and scavenger hunt area

Copyright © 2006 SafetyFUNdamentals™

Instructions for Scavenger Hunt

Organize the class into teams of 3-5. Present each team with a list of items they must collect and bring back to the class within the allotted time frame. Explain the rules clearly before they leave the training room. At least one of these rules should be NO RUNNING or otherwise UNSAFE ACTS are to be permitted in collecting any of the items. Another rule should be that the team members must stay together. You can prepare a list (ten items are good) to reflect the class content or overall safety principle. Items should be listed in a descriptive fashion - not just listed. For example, you could list "Find a glove that could be worn with acids" instead of just listing "glove." Another example could be "Find and copy an MSDS for a flammable liquid" instead of just listing "MSDS".

When the class returns with their findings go through the Scavenger Hunt list and ask each team to hold up their item that corresponds to the list.

Note: If possible, develop a separate list for each team so that the same items are listed but in a different order so that all teams are not going for the same items at the same time.

Debriefing/Discussion Questions:

What was the hardest thing on the list to find and why?

What strategy did you use to collect all of the items as quickly (and safely) as possible?

What item directly relates to an activity you will do back on the job this week?

Copyright © 2006 SafetyFUNdamentals™

Scrambled Eggs

Objectives	To share questions with classmates
Good For	Smaller Classes
Cautions	Reading and writing required
Class Length	Any length
Audience	Any
Where it can be done	Anywhere
Team Size	No teams - individual exercise
Time Needed	10 - 15 minutes
Set-Up/Materials	Plastic egg (like those sold at Easter time) for each trainee plus small slips of paper

Copyright © 2006 SafetyFUNdamentals™

Instructions for Scrambled Eggs

Give each trainee a small piece of paper and a plastic egg. Ask them to write a question about the topic on the paper and put it inside their plastic egg. Collect all of the eggs and "scramble" them, i.e., put the eggs into a basket and mix them up. Give each trainee an egg. Tell them to open the eggs and try to answer their question. If they do not know the answer, they can close their egg up and trade with someone else. (If someone asks them to switch, they must say yes.) After everyone has a question they know the answer to, go around the room and ask each trainee to read their question and related answer.

Debriefing/Discussion Questions:

How did you decide if you should switch your question or not?

How did it make you feel to have to answer your question in front of the class?

Why did you put the question that you did into your egg?

Copyright © 2006 SafetyFUNdamentals™

Six Degrees of Safety

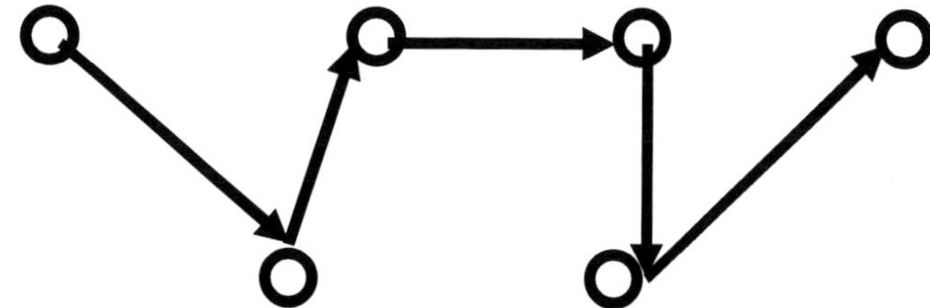

Objectives	To make connections between seemingly unrelated objects or tasks as a way to review relevant safety information
Good For	Classes with a lot of content
Cautions	More difficult than other activities. Writing required.
Class Length	Any
Audience	Trainees willing to think creatively
Where it can be done	Anywhere
Team Size	3-6
Time Needed	15-20 minutes
Set-Up/Materials	Paper and pens or pencils

Copyright © 2006 SafetyFUNdamentals™

Instructions for Six Degrees of Safety

Ask if anyone is familiar with the phrase "6 degrees of separation" and if so, ask them to tell the class what this expression means. If no one has heard the term, explain the concept (Note: 6 degrees of separation is the idea that everyone is connected to everyone else in 6 steps or less). Divide the class into teams. Assign each team a starting word or phrase and a finishing word or phrase. Explain that they have to think of the 4 additional steps in between the first and the last to get to the final word or phrase. Share the following example:

"Connect Forklift to Respirator"

While this might seem strange and impossible, trainees will need to recall a great deal of what they already know about safety in a variety of areas in order to make these connections. For the above example, one solution could be:

1) Forklift connects to 2) seatbelt which connects to 3)adjustable seating which connects to 4)one size does not fit all which connects to 5)fit testing which connects to 6)respirator

Encourage trainees to use their imagination, experience and anything related to workplace safety and health that they can remember.

When each team has finished, go around the room to each team and ask to hear their connections.

Debriefing/Discussion Questions:

Did you find this exercise easy or hard and why?

How did this help you to remember what you have been told about safety in the past?

Copyright © 2006 SafetyFUNdamentals™

Spin The Test Tube

Objectives	To encourage all trainees to participate in the learning experience
Good For	A preview or review of any safety topic
Cautions	In very large teams, all trainees will not be able to participate
Class Length	Any
Audience	Any
Where it can be done	Anywhere but the class needs to get into a circle and a flat surface is needed
Team Size	Twelve or less is best
Time Needed	About 1-2 minutes per trainee
Set-Up/Materials	One test tube (or other safety item that spins)

Copyright © 2006 SafetyFUNdamentals™

Instructions for Spin the Test Tube

First tell the class that this is not like any other spin the bottle game they may have played when they were younger. In this game, you will spin the bottle first and whoever it points to must either ask a question related to the topic or state one thing they know about the topic. If they ask a question, anyone in the circle can answer. After the question is asked or a statement is made, the bottle is spun by the first participant and whoever it points to next must do the same. This continues until everyone had had a chance to go.

This activity can be used as an opening to a class that is taught every year if you believe that most trainees are fairly knowledgeable about the material. It will also highlight what areas you need to cover in depth and which areas only need a brief review. If the activity is completed at the end of the class, it can highlight areas that need additional discussion as well as serve as an overall review of the class material.

Debriefing/Discussion Questions:

Why would you rather ask a question or state something that you know?

What did you hear during the exercise that you can use back at work?

What was the most important thing that you heard?

Copyright © 2006 SafetyFUNdamentals™

Still Standing

Objectives	To review class information
Good For	Opening annual refresher classes or reviews
Cautions	Tossable objects should be soft
Class Length	Any
Audience	Any
Where it can be done	Anywhere
Team Size	2
Time Needed	Depending on class size, approximately 10 minutes
Set-Up/Materials	A tossable object like a Koosh® ball for ½ the class, prepared questions and music

Copyright © 2006 SafetyFUNdamentals™

Instructions for Still Standing

Before the class, write out a number of fairly difficult questions that you will read to the class during the exercise.

Have the trainees stand in two lines facing each other. Give each pair a tossable object like a Koosh® ball. If you can find tossable objects related to the class topic, such as an inflated rubber glove if the topic is hand safety, you could use that as well.

Have the two lines stand about 3 - 5 feet apart. Tell the class that when the music starts they should start passing the object back and forth. When the music stops, whoever has the object must answer the question that you ask of the group. If they get it right, then the other person sits down. If they get it wrong or do not know it, they sit down. After the first round, individuals are paired up with new partners made up from the trainees still standing. The music starts again and is once again stopped. The second question is asked and the same procedure is followed. This continues until there are two trainees left standing. The music is played for the final time and the last question is asked. The trainee left standing is declared the winner.

Debriefing/Discussion Questions:

How did it feel having pressure added to the requirement to answer each question?

How could this effect safety in the workplace?

Copyright © 2006 SafetyFUNdamentals™

Stupid People Tricks

Objectives	To provide examples of case studies
Good For	Topics where the class has a high level of experience with the subject matter
Cautions	People that others know should not be described in the exercise
Class Length	At least an hour
Audience	Any with a high level of industrial experience
Where it can be done	Anywhere
Team Size	Individual activity
Time Needed	At least 15 minutes
Set-Up/Materials	Inform trainees *in advance* that they will need to share a story-related to the topic

Copyright © 2006 SafetyFUNdamentals™

Instructions for Stupid People Tricks

Inform the trainees in advance of the class that they will be asked to share a story related to safety that would classify as a "stupid person trick." They can be prepared to come and tell a story about someone they personally know, a story they heard from someone else or even one they read about. Warn the trainees that they are not to bring a "stupid person trick" that is directly related to someone that others know. Tell them that they are also not to share the names of the individuals involved.

When they arrive to the class, ask for volunteers to begin sharing their "Stupid People Tricks." Write the name of each "trick" on a flipchart or board. After everyone has had a chance to go, tell the trainees you will now vote on the stupidest trick. Each trainee can only vote once and they cannot vote for their own story. In the end, announce the winner as the individual who shared the story that received the most votes.

Debriefing/Discussion Questions:

What stupid people trick could occur in your workplace and why?

How can this be prevented from occurring?

What lesson did you learn from one of the "Stupid People Tricks" you heard?

Copyright © 2006 SafetyFUNdamentals™

Take a Spin

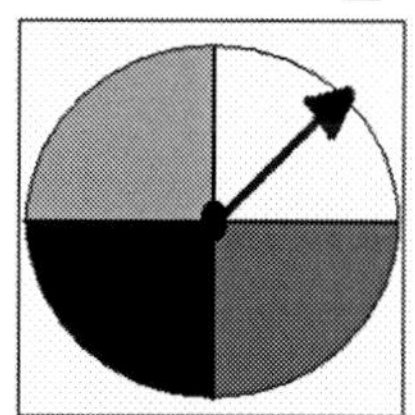

Objectives	Review of class information - either as a review for an annual refresher or as a summary review
Good For	Reviewing class information in four categories
Cautions	Reading required
Class Length	Any
Audience	Anyone
Where it can be done	Anywhere
Team Size	3-5
Time Needed	15-20 minutes
Set-Up/Materials	Four color spin wheels (either purchased or made), questions for each corresponding color, paper for keeping score

Copyright © 2006 SafetyFUNdamentals™

Instructions for Take a Spin

Before class, decide on the four categories of information you want to cover and assign each of these categories a color on the spinner. Next, write up approximately 10 questions for each category and place these questions on corresponding cards of the same color. (Note: Index cards come in a wide variety of colors. Consider typing the questions on the computer and printing out several sets of labels. The labels can then be applied to the index cards). Answers should not be provided on the question cards.

Divide the class into groups of 3 - 5. Give each group a spinner and a set of cards. The cards should be placed in 4 piles by color, and left face down on the table. Each person in the group should spin the wheel, and then answer the corresponding question on the card. If they get it right, they get one point. If they get it wrong, nothing happens. Note: the correct answers are not provided on the question cards so to determine if an answer given was correct the other members of the group should discuss the question and decide if the point should be awarded. The spinner is passed to the person seated to their left for their turn. All players continue to spin and answer until all of the questions have been answered.

When each team has finished, ask for the team's total points - not each individuals. Announce the team with the most points as winner. Note: do not tell the class ahead of time that you will be asking for the final points tally in this way.

Debriefing/Discussion Questions:

How did your group come to consensus on each correct answer?

What was the most difficult question and why?

What information did you find most useful and why?

Copyright © 2006 SafetyFUNdamentals™

Talk and Toss

Objectives	To encourage brainstorming
Good For	Topics where you want to develop new ideas
Cautions	Use a soft ball
Class Length	Any
Audience	Any
Where it can be done	Anywhere
Team Size	None needed for smaller classes. In very large classes, the class can be put into teams of 8 -10
Time Needed	About 10 minutes
Set-Up/Materials	One soft ball

Copyright © 2006 SafetyFUNdamentals™

Instructions for Talk and Toss

Place the class into groups of 8 - 10. (For smaller classes, the entire class can be one group). Have them stand at their seats or in a circle and toss the ball to one person. Tell the class the question they will be brainstorming. The question should be appropriate to the class and related to an area where you would like to encourage trainees to develop new ideas. For example, in a class where you are training supervisors on providing positive feedback to their employees, you could make the question "What are different ways you can tell an employee they are doing a good job?" Once you ask the question, the individual with the ball tosses it to someone else who must give their answer and then toss it to another person and so on. After the first person tosses the ball, he or she becomes "scribe" and should start writing down as many of the answers that they hear as quickly as they can. (If they miss some, that is okay). After everyone has gone, ask the scribe to read back some of the answers.

Debriefing/Discussion Questions:

What ideas did you hear that you would not have thought of on your own?

How will you use this idea and when will you try it?

Copyright © 2006 SafetyFUNdamentals™

Test and Check

Objectives	To encourage trainees to remember difficult class information
Good For	Any topic
Cautions	Make sure all questions are fair; reading and writing is required
Class Length	At least 45 minutes
Audience	Any
Where it can be done	Anywhere but a classroom would be best
Team Size	3-5
Time Needed	At least 15 - 20 minutes
Set-Up/Materials	Large index cards, pens or pencils

Copyright © 2006 SafetyFUNdamentals™

Instructions for Test and Check

After content is presented or at the beginning of an annual refresher class, divide class into teams of 3-5. Give each team a large index card (and an extra in case of mistakes). Tell each team that their goal is to think of 3 questions related to the class content that they do not believe the other teams will be able to answer. They can make the questions as difficult as possible but they must be fair questions and the answers must be covered in the class material. There are no "trick questions" allowed and notes and hand-outs cannot be used in writing or in answering the questions.

After each team has had time to come up with 3 questions, each team should pass their question card to the team to their right. Each team should be given 5-10 minutes to come up with the best answers for those three questions. For each question each team is able to answer correctly, they get 1 point. For every question they are unable to answer, the team that wrote the question gets a point.

When all teams are finished, ask each team to report their questions and their answers. Keep track of all points on a flip chart and award a small prize to the winning team.

Debriefing/Discussion Questions:

Why did you think the questions you submitted would be difficult to answer?

What was the most difficult question you had to answer and why?

Was there one particular question where your team had difficulty agreeing on an answer and if so, how did you finally agree?

Copyright © 2006 SafetyFUNdamentals™

Theater Teams

Objectives	To demonstrate a thorough understanding of a process or task
Good For	Safety-related work assignments
Cautions	Many people are uncomfortable "acting"
Class Length	At least 30 minutes
Audience	Any
Where it can be done	Anywhere
Team Size	3-5
Time Needed	Approximately 15 - 20 minutes
Set-Up/Materials	Prepared "scene" cards for each team and a timer

Copyright © 2006 SafetyFUNdamentals™

Instructions for Theater Teams

Place the class into acting troupes (teams). Tell them they will need to act out a particular aspect of a safety task related to the class content. All members of each team should be involved in the performance. The rest of the class will have to guess what they are doing and the team that gets their message across the fastest, wins. You can give the teams the option of thinking up their own task to act out or you can let them pick a "scene" card (which you prepare before the class) to find an appropriate activity. For example, in a class on lock-out/tag-out, the scene cards could include the process of locking out at a panel, writing a tag and placing it on the machine, notifying others working in the area or even receiving a shock from not following LOTO procedures.

After each team has had adequate time to prepare their act, have each team go to the front of the class and demonstrate their activity. Tell the rest of the class to shout out the name of the activity once they think they know the answer. When the teams begin, start the timer and then keep track how of how long it takes each team to get their message across to the rest of the class. Award a prize to the team that gets their message across the fastest. (You might not want to announce that you are keeping track until you award the winner - it could cause some very competitive teams to withhold guessing correctly so that the performing team's time is increased.)

Debriefing/Discussion Questions:

What was the most difficult thing about this activity?

What activity was acted out that applies directly to you and your job?

Were the activities all done safely and if not, what should have been done differently?

Copyright © 2006 SafetyFUNdamentals™

Training Tutor

Objectives	To encourage the class to take an active role in sharing information and training their co-workers
Good For	Annual refreshers or other topics where trainees are very familiar with class content
Cautions	Trainees may not remember as much as they think they do
Class Length	At least 60 minutes
Audience	Any
Where it can be done	Anywhere
Team Size	2-5
Time Needed	Approximately 40 minutes
Set-Up/Materials	Nothing

Copyright © 2006 SafetyFUNdamentals™

Instructions for Training Tutor

Tell the class that you need help designing a new training class on a particular topic. Divide the class into five teams of 2 - 5 members per team. Assign one team the introduction, one team the middle, one team the summary, one team the quiz and one team the activity. Tell the class that they have 20 minutes to work on their section of the training. They should write down as many details and ideas as they can for their section of the training. When they are done, call on each team individually and in order to share their plan with the class. Be prepared to help each group with ideas for their section.

Debriefing/Discussion Questions:

Did you find this activity easy or hard and why?

What information did you forget from previous classes?

What do you feel is the most important piece of information to be conveyed to the class?

Copyright © 2006 SafetyFUNdamentals™

Tree Branches

Objectives	To help the trainee to develop a memory aid, similar to a Mind Map, to remember class content
Good For	Classes with a lot of material
Cautions	Writing required
Class Length	Any but longer classes preferred
Audience	Anyone
Where it can be done	Anywhere
Team Size	Any
Time Needed	15-20 minutes
Set-Up/Materials	Blank paper (or pre-printed tree papers) and pen or pencils

Copyright © 2006 SafetyFUNdamentals™

Instructions for Tree Branches

This activity can either be explained to the trainees before you start reviewing content with the instruction to work on their tree as the class goes on or the activity can be introduced at the end of the class as long as you give the trainees adequate time to work on filling in the branches of their tree with the class information. For a 5-branch tree as shown in Appendix L, it is sometimes useful to write five main training ideas on each main branch before starting.

The Tree Branch activity works very similar to a Mind Map. It is a visual aid to help trainees to visually organize what they learned and to help them recall the information in the future. Explain that when they hear an idea or a concept related to one of the five main ideas, they should write it on the appropriate branch. They should then write details that are connected to that idea on the smaller branches. At the end of the class, ask for a volunteer to share his or her tree with the class, explaining the main and detail branches.

Debriefing/Discussion Questions:

How is your tree different from your co-workers' trees?

How will the tree help you to remember the information presented in class?

What other key ideas would you list on the tree?

Copyright © 2006 SafetyFUNdamentals™

Watercolor Work

Objectives	To review class information
Good For	Reviews - before, during, or at the conclusion of the class
Cautions	None
Class Length	At least 45 minutes
Audience	Any
Where it can be done	Classroom
Team Size	4
Time Needed	At least 10-15 minutes
Set-Up/Materials	Watercolor paint box and paintbrush for each trainee, small cups of water, paper and a set of four keywords for each team

Copyright © 2006 SafetyFUNdamentals™

Instructions for Watercolor Work

Before class, write out 4 key phrases related to the class topic on four separate cards. Each phrase should be written on a separate card. A set of the 4 cards will be needed for each group.

Divide the class up into teams of 4. Give each team a set of watercolor paints, a small cup of water, and a paint brush and paper for each person in the group. Go to each group and let each trainee pick a card out of an envelope. They are not to show their card to their team members. (Note: The same 4 phrases can be written on each team's set of cards since the teams will not be mixing during this activity). Each trainee, without showing the others in his or her group, will need to paint a picture of the phrase they selected. They should have about five minutes for this exercise. After 5 minutes, they should stop. Each trainee should put their painting in the center of the table side by side and the other team members must try to guess what phrase or activity each watercolor drawing depicts. The painter should not reveal what the painting is until you have given the class adequate time to guess (and laugh!) When guessing, the trainees will have to mentally review everything they learned.

Note: the phrases should be somewhat complex and require some thought to paint. For example, in a class on hot work, a phrase such as "remove or cover all combustibles" would be a great phrase to have to try to paint.

Debriefing/Discussion Questions:

How do you feel about doing an activity such as this that you probably haven't done in a very long time? Why?

How did trying to guess what the other paintings were help you to remember the key points presented in this class?

Copyright © 2006 SafetyFUNdamentals™

Who am I?

Objectives	To review safety responsibilities of a diverse population
Good For	Safety programs or processes that involve many departments
Cautions	Reading required
Class Length	At least 45 minutes
Audience	Any
Where it can be done	Anywhere
Team Size	At least the number of trainees as "personalities"
Time Needed	10 - 15 minutes
Set-Up/Materials	"Personality" card for each trainee and timer

Copyright © 2006 SafetyFUNdamentals™

Instructions for Who Am I?

Before class, write out a personality card for each trainee.
Either after the presentation of the material is complete or before a class starts, give each trainee one of the cards. Tell the class that each of the cards describes a role that someone in the facility has with respect to the topic. (For example, for a class on accident investigation, the cards would represent the roles of the supervisor, the injured employee, the medical personnel, the safety manager, a witness, etc.) Each individual will have one minute to say things or do things that the individual on their personality card would have responsibility for as related to the topic. After 1 minute, stop the timer and ask the other class members to write down who they think the "actor" was portraying. Review the personality. For every correct guess the actor gets 1 point. After everyone in the class has had a chance to act out his/her "personality," announce the winner based on the number of points each person received.

Debriefing/Discussion Questions:

Why was it easy or difficult to guess the "personality"?

What, if anything, did the actors leave out that should have been included?

How can this activity help you to understand the challenges faced by each "personality?"

Copyright © 2006 SafetyFUNdamentals™

Write-On BINGO

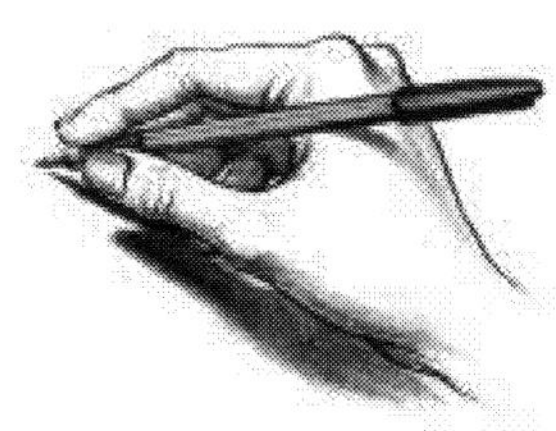

Objectives	Encouraging trainees to work together to apply class knowledge to their daily tasks
Good For	Any topic
Cautions	Reading and writing required
Class Length	At least 45 minutes
Audience	Any
Where it can be done	Anywhere
Team Size	2-3
Time Needed	Depending on the questions, 10-20 minutes
Set-Up/Materials	Prepared Write-On Bingo Sheets for each team, pens or pencils, timer

Copyright © 2006 SafetyFUNdamentals™

Instructions for Write-On BINGO

Divide the class into teams of 2 or 3. If possible, individuals from the same work areas should be on the same teams. Give each class a prepared Write-On BINGO sheet. In Write-On BINGO, each block has a question or clue. The team must write the answer to that question or clue in the block in order to be able to consider it marked off. Questions should be challenging and require the teams to think about their work responsibilities specifically. (This also means that there will not be a right or wrong answer). For example, one square of the Write-On BINGO sheet could say "name one corrosive you use in your work area." The first team to mark off an "X" or a "+" or a "T" shape (any orientation) wins. If you have more time, you could also make the winner the first team to completely cover their BINGO sheet. Another alternative would also be to make different BINGO sheets for different teams.

After the winning team has finished, have them present their winning sheet to the class by reading the questions they answered and the answers they provided.

A sample Write-On BINGO sheet is provided in Appendix M. To download a blank write-on BINGO template, visit www.crownsafety.com.

Debriefing/Discussion Questions:

What was the most difficult question to answer and why?

What question was the easiest and why?

How did working with others from your work area help you?

Copyright © 2006 SafetyFUNdamentals™

Part 3

APPENDICES

Note:

Samples found in the Appendices are also available electronically at www.CrownSafety.com.

When attempting to access this free information available only to book purchasers, follow the SafetyFUNdamentals Club link and use the ID "Funtrainer" and the Password "airplane."

Please do not share this ID and Password with others.

Copyright © 2006 SafetyFUNdamentals™

Copyright © 2006 SafetyFUNdamentals™

APPENDIX A

Example BINGO Game for an Office Ergonomics Training Class

Note: Only one BINGO Card is shown here. To effectively use a BINGO game, it is preferable that each trainee have a unique card. To download a full BINGO training game for a class of up to twenty trainees, visit the SafetyFUNdamentals™ Book Club at www.crownsafety.com.

Office Ergonomics

glare	monitor	mouse	wrist rest	awkward posture
repetitive motion	Carpal Tunnel Syndrome	tendonitis	eye strain	flat on floor
elbow height	close to body	knee height	shoulder height	lower back
right angles	arm's reach	twisting	contact stress	neck pain
headaches	task lighting	eye level	force	breaks

Copyright © 2006 SafetyFUNdamentals™

Copyright © 2006 SafetyFUNdamentals™

"Call Clues" for Office Ergonomics Bingo Game

Note: the answers are shown in parentheses

1. When direct light hits the face of a monitor, this can occur (glare)

2. When you stop typing, this gives your wrist a place to rest (wrist rest)

3. If the font on your screen is too small or hard to read, this can result (eye strain)

4. Lifting an object while also doing this can lead to an injury (twisting)

5. When working on a computer, this part should be directly in front of you (monitor)

6. When sitting at a desk, your elbows and arms and shins and thighs should be in this position (right angles)

7. When sitting at a computer station, your monitor should be approximately this far away (arm's reach)

8. A repetitive motion injury that results from too much pressure in the wrist (Carpal Tunnel Syndrome)

9. When storing heavy boxes and supplies, it is best if they are stored above ____________ (knee height)

10. Squinting at a computer for long hours can cause these (headaches)

11. This should be used while keeping your arm close to the body (mouse)

12. Typing with your wrist bent up or down is considered this (awkward posture)

13. Activities at work or home that involve doing the same thing over and over are an example of ___________ (repetitive motion)

Copyright © 2006 SafetyFUNdamentals™

Copyright © 2006 SafetyFUNdamentals™

"Call Clues" for Office Ergonomics Bingo Game Continued......

14. When a tendon becomes inflamed from overuse it is called ___________ (tendonitis)

15. When sitting at a computer workstation typing, your feet should remain ___________________ (flat on the floor)

16. When lifting any heavy load, the load should be kept _________ (close to the body)

17. When storing items on shelves, you should try not to place heavy items above _____________ (shoulder height)

18. When sitting in a chair for an extended period of time, it is important to support this (lower back)

19. If you are working with your wrists or forearms resting or leaning against the square edge of a desk or table, you could experience this (contact stress)

20. If you repeatedly hold a phone receiver or cell phone between your ear and shoulder, you could experience this (neck pain)

21. If glare is a problem from overhead lights, this type of lighting may work better for you (task lighting)

22. The top of your computer screen should be approximately an inch or two below this (eye level)

23. Injuries can occur when using too much of this when hitting the keys (force)

24. A great way to prevent aches and pains associated with sitting at the computer too long is to take these (breaks)

25. When typing at a keyboard, the keyboard should be at approximately this level (elbow height)

Copyright © 2006 SafetyFUNdamentals™

Copyright © 2006 SafetyFUNdamentals™

APPPENDIX B
Clockwork Diagram

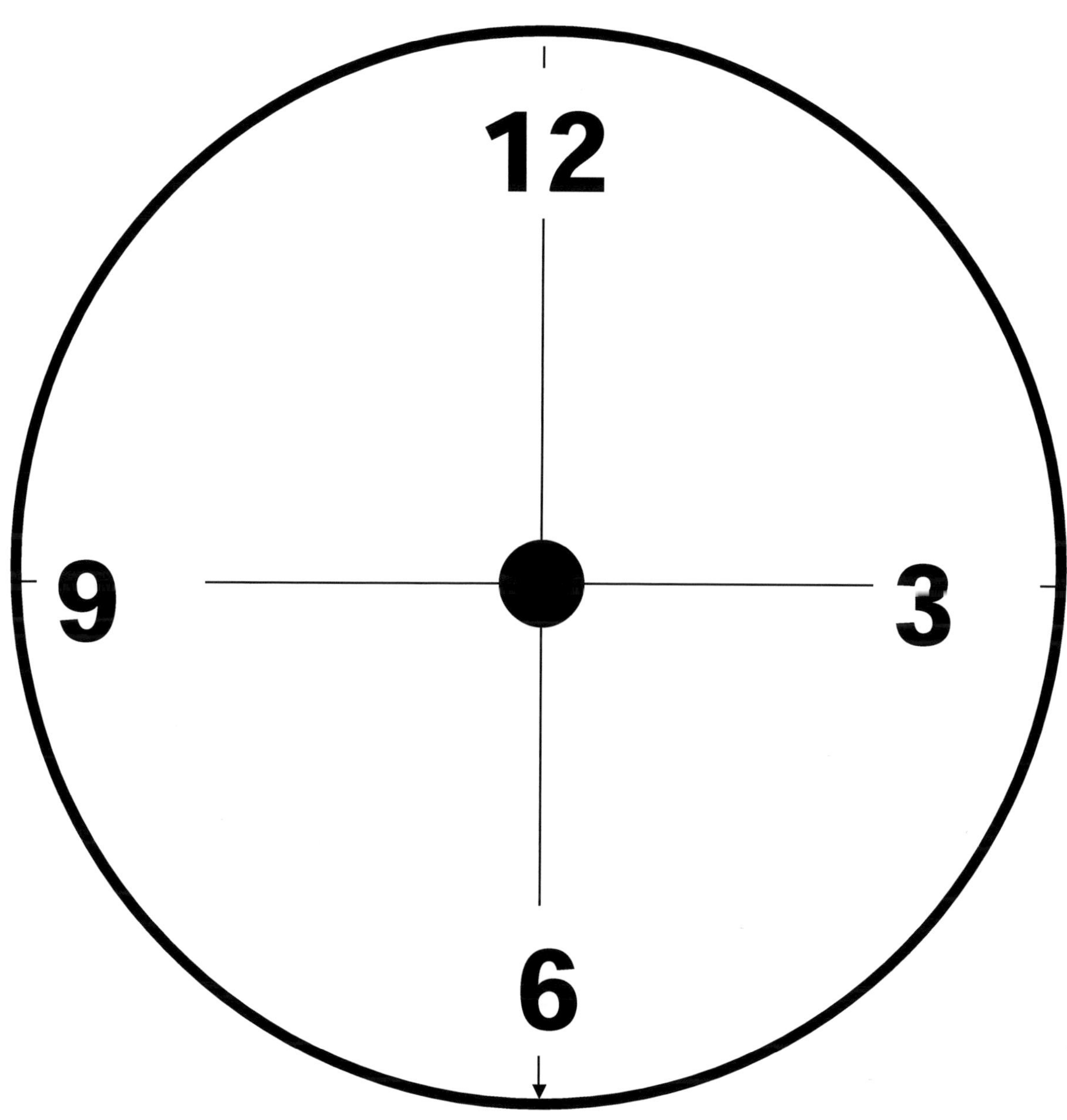

Copyright © 2006 SafetyFUNdamentals™

Copyright © 2006 SafetyFUNdamentals™

safetyFUNdamentals

APPPENDIX C
Crazy Crossword Blank

Across

5.
6.
9.
10.
11.

Down

1.
2.
3.
4.
7.
6.

Copyright © 2006 SafetyFUNdamentals™

Copyright © 2006 SafetyFUNdamentals™

APPPENDIX D
Cryptogram for a Housekeeping Class

B G G V   E G Y I L C L L S Z H B   D P H   E L F S

S X L O L H Q   I F Z S I ,   Q X Z S I   P H V   N P F F I

Z H   Q E L   A G X C S F P D L .

Answer:

Good Housekeeping Can Help Prevent Slips, Trips and Falls in the Workplace.

Copyright © 2006 SafetyFUNdamentals™

Copyright © 2006 SafetyFUNdamentals™

APPPENDIX E
Four Square Diagram

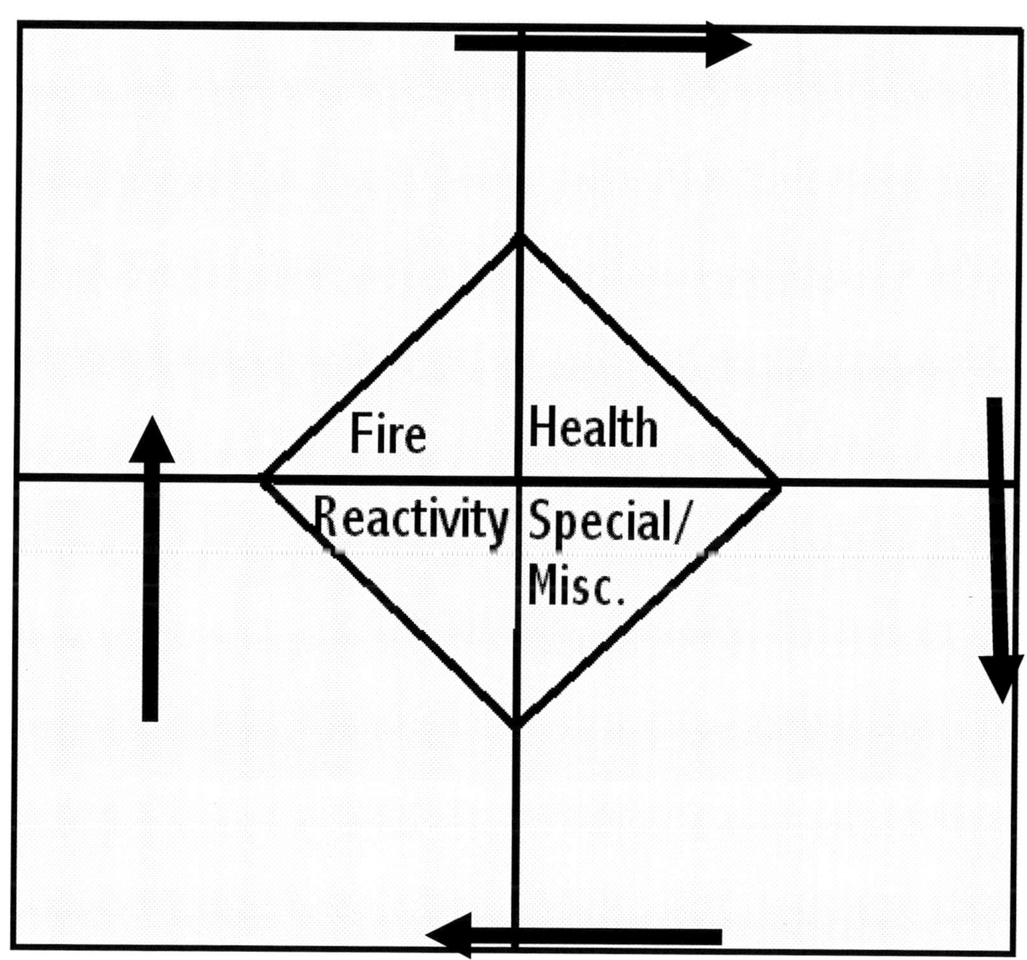

Copyright © 2006 SafetyFUNdamentals™

Copyright © 2006 SafetyFUNdamentals™

APPENDIX F
Forklift Frenzy Game Board Set Up

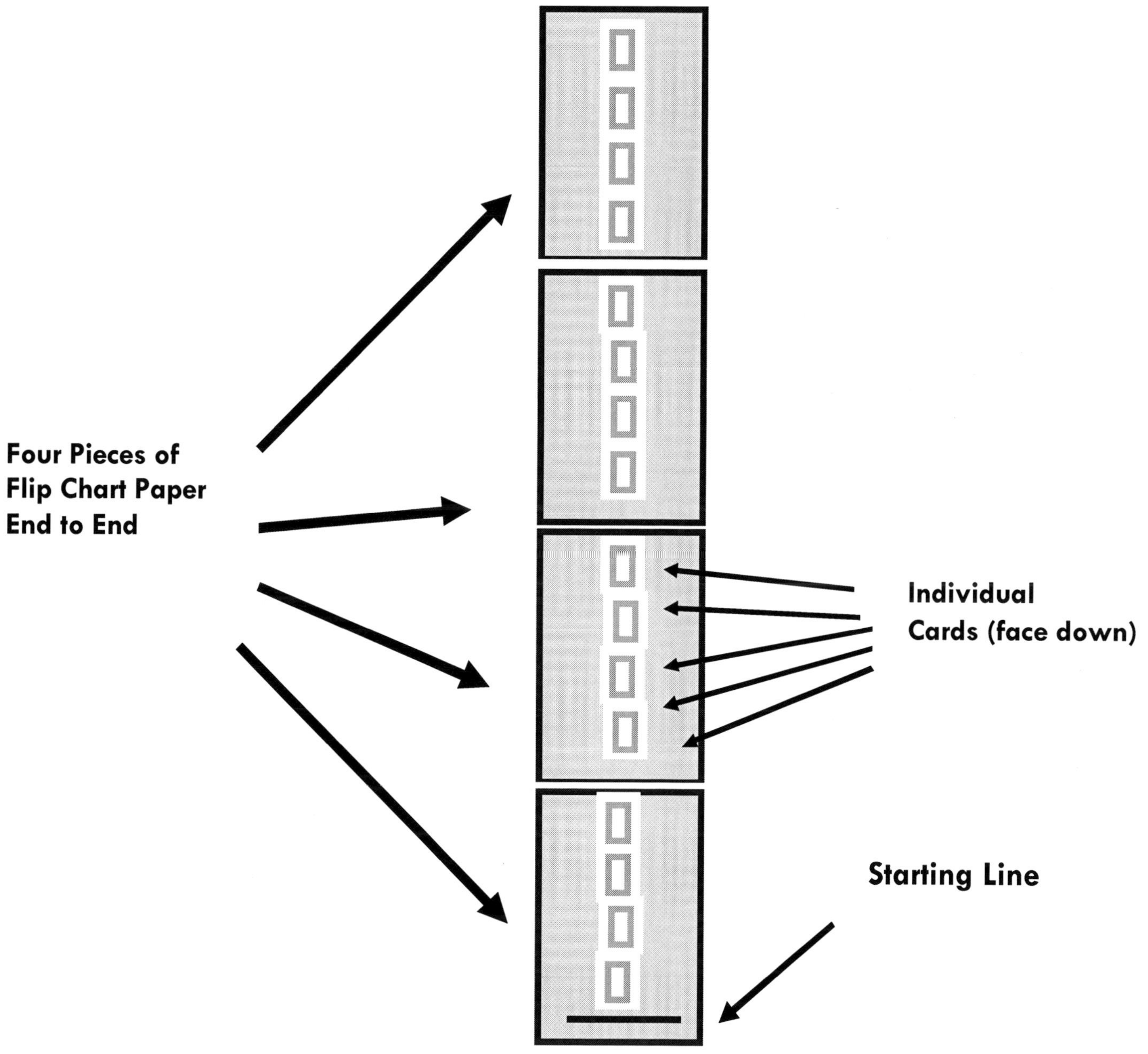

Copyright © 2006 SafetyFUNdamentals™

Copyright © 2006 SafetyFUNdamentals™

APPPENDIX G
HazCom Hill Game Board

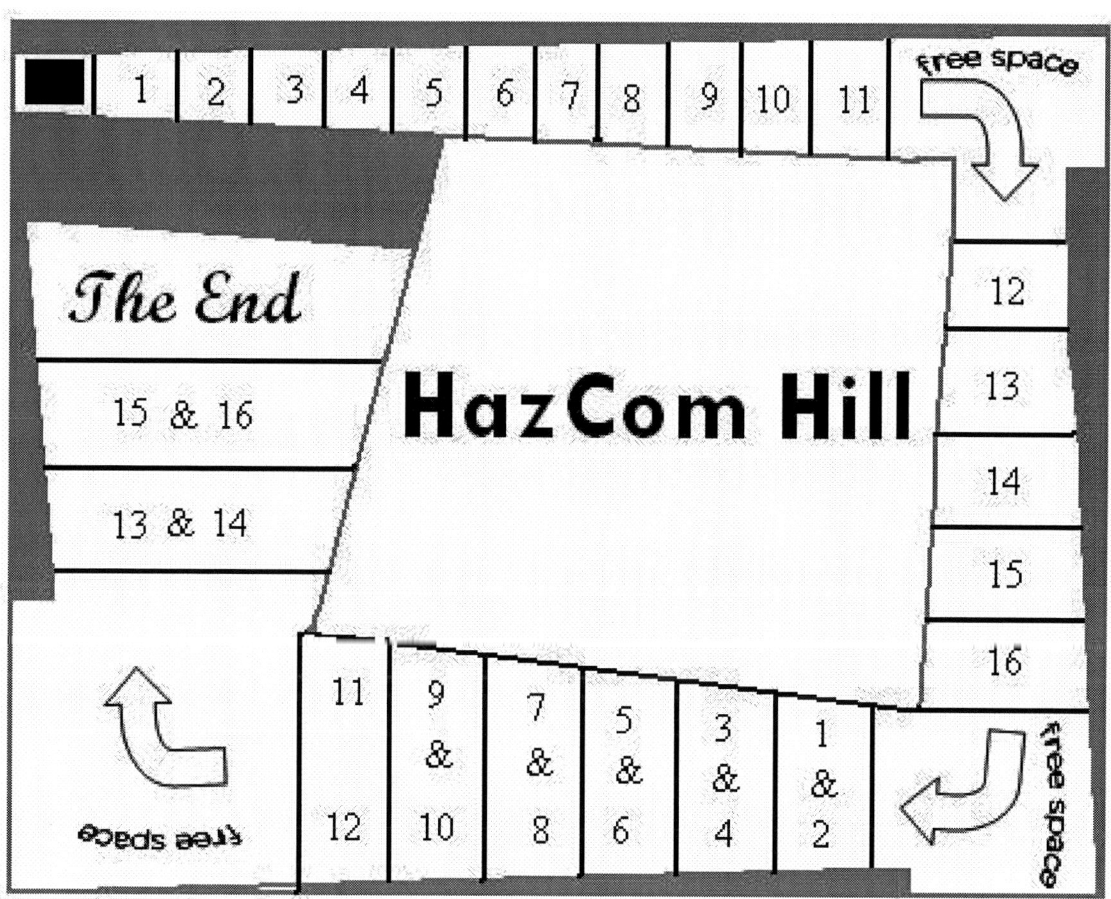

Copyright © 2006 SafetyFUNdamentals™

Copyright © 2006 SafetyFUNdamentals™

APPPENDIX H
Plot the PPE Matrix

	P	L	O	T
Category 1				
Category 2				
Category 3				

Copyright © 2006 SafetyFUNdamentals™

Copyright © 2006 SafetyFUNdamentals™

APPPENDIX I
Safety Hangman

There are 8 chances to guess the letters of the mystery word. For every wrong guess, cross off one of the following pieces of PPE: hardhat; ear muffs, safety glasses, vest, left glove, right glove, left shoe and right shoe. When there is no PPE left, the game is over.

Copyright © 2006 SafetyFUNdamentals™

Copyright © 2006 SafetyFUNdamentals™

APPENDIX J
Safety-Lib Example

There was an employee that worked in the ______________
(name of company department)

department of our company. One day, before he came to work,

he ______________ out of his car and ___________ across the
(verb/action word) (verb/action word)

_________________. He slipped on a ______________ and
(place/area of company) (noun/object)

broke his ______________ and ______________________.
(body part) (body part)

Next time, he will be more careful and will not ______________
(verb/action word

or ______________ any more. Instead, he will ask __________
(verb/action word) (name of someone)

to ______________ for him. To prevent this accident from
(verb/action word)

occurring _________he will always wear his _______________
(point in time) (type of PPE)

when he _________.
(verb/action word

Copyright © 2006 SafetyFUNdamentals™

Copyright © 2006 SafetyFUNdamentals™

APPENDIX K

Safety Menu Example

Starters

Safety Glasses	$5/pair
Safety Goggles	$5/pair
Hardhats	$10 each
Safety Vest	$10/each
Caution Tape	$3/roll
L.O.T.O. Tags	.50/each
L.O.T.O. Locks	$5/lock

Main Course

Tripod	$30
Eyewash	$50
Safety Shower	$75
Machine Guard	$50
Light Curtain	$200
Fire Extinguisher	$30
Intercom System	$200
Drum Handler	$200
Spill Clean Up Kit	$200
Portable exhaust	$350
Walkie Talkies	$100 for two

Desserts

First Aid Kit	$20
Gatorade	$5/gallon
Polaroid Camera	$20
Deicing Salt	$5/bag
Suntan Lotion	$5/bottle

Copyright © 2006 SafetyFUNdamentals™

Copyright © 2006 SafetyFUNdamentals™

APPENDIX L
Tree Branches Sample Diagram

Copyright © 2006 SafetyFUNdamentals™

Copyright © 2006 SafetyFUNdamentals™

APPENDIX M
Write-On BINGO Example

A corrosive In your work area	An acid in your work area	The pH of something you work with	The location of the nearest fire extinguisher	The location of the nearest eyewash
Location of nearest safety shower	Location of MSDS	Location of first aid kit	Location of 3 emergency stops in your area	Location of fall protection equipment
Location of nearest spill kit	Eye Protection area	Hearing protection area	Location of bloodborne pathogens clean up kit	Closest phone to call 911
Number to dial to report an emergency	Location to get new work gloves	Location to get new safety shoes	3 things to remember about fire hazards	Location of nearest sprinkler
Location of nearest flammable storage area	Location of temporary storage of hazardous waste	Location of nearest water fountain	Location of closest emergency exit	Emergency meeting spot outside of building

Copyright © 2006 SafetyFUNdamentals™

Copyright © 2006 SafetyFUNdamentals™

Part 4

Suggested Reading

Order Forms

Let's Hear It!

About the Author

Copyright © 2006 SafetyFUNdamentals™

Copyright © 2006 SafetyFUNdamentals™

Suggested Reading

Biech, E., *Training for Dummies*. Hoboken, Wiley, 2005

Furjanic, S. and L. Trotman. *Turning Training Into Learning*, New York, Amacom, 2000.

Meier, D. *The Accelerated Learning Handbook*, New York: McGraw Hill, 2000.

Pike, R., Creative *Training Techniques Handbook*. Minneapolis, Lakewood Books, 1994.

Rose, E., *Presenting and Training with Magic*. New York, McGraw Hill, 1998.

Ruddell, L. *The Accelerated Learning Fieldbook*, San Francisco, Jossey-Bass/Pfeiffer, 1999.

Stolovitch, H. and E. Keeps. *Telling Ain't Training*, Alexandria, ASTD, 2002.

Thiagarajan, S. *Design Your Own Games and Activities*, San Francisco, John Wiley & Sons, 2003.

Wacker , M. and L. Silverman. *Stories Trainers Tell*, San Francisco, Pfeiffer, 2003

Callahan, M.*10 Great Games and How to Use Them*, Alexandria, VA, ASTD, 1999

Fairbanks, D. *Accelerated Learning*, Alexandra, VA, ASTD, 1997

Furjanic, S. and L. Trotman. *Turning Training Into Learning*, New York, Amacom, 2000.

•

Copyright © 2006 SafetyFUNdamentals™

Copyright © 2006 SafetyFUNdamentals™

Suggested Reading Continued...

Meier, D. *The Accelerated Learning Handbook*, New York: McGraw Hill, 2000

Pike, R. Creative *Training Techniques Handbook*. Minneapolis, Lakewood Books, 1994.

Rose, E. *Presenting and Training with Magic*. New York, McGraw Hill, 1998.

Ruddell, L. *The Accelerated Learning Fieldbook*, San Francisco, Jossey-Bass/Pfeiffer, 1999.

Silberman, M. *Active Learning: 101 Strategies to Teach Any Subject*, Needham Heights, Allyn & Bacon, 1996

Stolovitch, H. and E. Keeps. *Telling Ain't Training*, Alexandria, ASTD, 2002.

Sugar, S. *More Great Games*, Alexandria, VA, ASTD, 2000

Tapp, L. "Accelerated Learning for Safety Training," Proceedings of the ASSE PDC, 2005

Tapp, L., "It's All Fun and Games :Twenty Five Activities to Make Training Great", Proceedings of the ASSE PDC, 2006

Thiagarajan, S. *Design Your Own Games and Activities*, San Francisco, John Wiley & Sons, 2003.

Treasurer, B. *Right Risk :Ten Powerful Principles for Taking Giant Leaps with Your Life*, San Francisco, Berrett-Koehler, 2003

Wacker , M. and L. Silverman. *Stories Trainers Tell*, San Francisco, Pfeiffer, 2003

Copyright © 2006 SafetyFUNdamentals™

Copyright © 2006 SafetyFUNdamentals™

Order Form

Orders can be placed online at www.SafetyFUNdamentals.com or by completing this form and mailing it with your check made payable to "SafetyFUNdamentals™" to the address below.

	Quantity	Price Each	Total Price
Additional Copies of SafefyFUNdamentals: 77 Games and Activities to Make Training Fun		**$47.95**	
SafetyFUNdamentals Workbook and Activity Kit Vol. 1 - everything you need for a class of twenty to do most of the games and activities described in this book		**$397.95**	
BINGO game for 25 on any safety or health topic (includes game cards and clue sheet)		**$29.95**	
Board Game Kit (blank game board, cards, dice, markers, spinner)		**$18.95**	
Using Games in Training 60 minute CD		**$17.95**	
Shipping is included on all orders	Orders to New Jersey include 7% Sales Tax		
	Total		

Mail Order Forms to:
SafetyFUNdamentals™
1999 Route 70 East, Suite 13
Cherry Hill, NJ 08003
OR Order Online at www.SafetyFUNdamentals.com

Copyright © 2006 SafetyFUNdamentals™

Copyright © 2006 SafetyFUNdamentals™

safetyFUNdamentals

Let's Hear It!

Do you have your own original safety training game or activity that you would like to share? Send it in with this form or email to MyIdea@safetyFUNdamentals.com and if we include it in our next book, we will send you a $25 gift certificate towards any training product offered on our website.

Your Name (required)____________________________

Company____________________________________

Phone: ______________________________________
(required - in case we have questions)

Email (Required)______________________________

Describe your game or activity on the back of this page, sign below, and you are finished. Don't forget to tell us all the details. Thanks for sharing your ideas!

- -

I agree that by submitting my training game idea for possible inclusion in a future publication that SafetyFUNdamentals™ may edit my submission and may use it in other media. I also understand that my submission may not be right for SafetyFUNdametals™ and if so, will not be used.

I agree:____

(Sign Your Name Here)

Copyright © 2006 SafetyFUNdamentals™

Copyright © 2006 SafetyFUNdamentals™

About the Author

Linda Tapp is President of SafetyFUNdamentals™, a safety training products company for safety trainers who care. She is also President of Crown Safety, a traditional safety, health and ergonomics consulting firm. Both companies are based in Cherry Hill, New Jersey. She received her Bachelor of Science in Biology from Drexel University and a Master of Science in Environmental Health from Temple University. She is a Certified Safety Professional (CSP) in Comprehensive Practice as well as in Ergonomics. She is frequently quoted and interviewed in the media and has appeared in USA Today, Fortune, Occupational Hazards, Occupational Safety, Safety Smart!, and Industrial Hygiene and Safety News. She frequently speaks at safety conferences across the country and has had over 50 articles published in professional and trade magazines.

In addition to "SafetyFUNdamentals™: 77 Games and Activities to Make Training Great," Linda has also written several eBooks on Safety Training Activities for specific topics and industries, including the free introductory eBook titled "Seven Keys to Using Games in Safety Training" which is available at www.GamesforSafety.com. She is currently working on "SafetyFUNdamentals™: More Great Safety Training Games" which is scheduled to be published in 2007.

Linda also provides a high energy, fully interactive, one-day workshop for serious safety trainers on all of the games and activities covered in the first SafetyFUNdamentals™ book. This workshop can also be provided on-site for corporations wishing to provide a group of safety trainers with the skills needed to effectively use games and activities in their training classes. For workshop information, visit www.SafetyFUNdamentals.com.

Copyright © 2006 SafetyFUNdamentals™

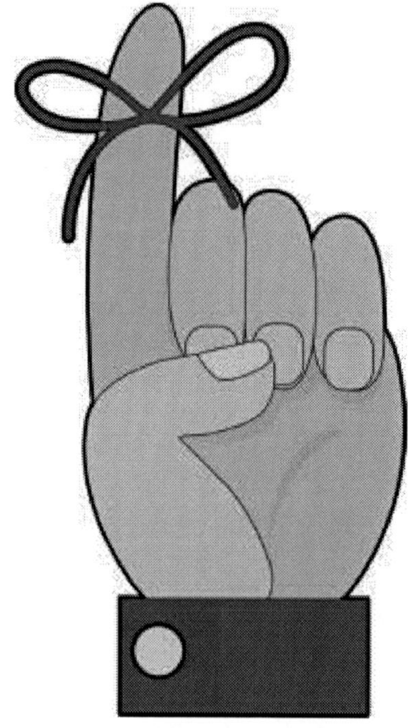

Don't forget to register your book!

If you purchased "SafetyFUNdamentals: 77 Games and Activities to Make Training Great!" anywhere but through the SafetyFUNdamentals website, we do not know who you are and will not be able to send you updates or our free bi-monthly Safety Training eZine which contains a free safety game in every issue plus valuable articles that can help you in your mission to provide great safety training classes.

Register Today at

www.SafetyFUNdamentals.com

Copyright © 2006 SafetyFUNdamentals™